# Energy Management

# Energy Management

## Conservation and Audits

Anil Kumar

Om Prakash

Prashant Singh Chauhan

Samsher

CRC Press
Taylor & Francis Group
Boca Raton London New York

CRC Press is an imprint of the
Taylor & Francis Group, an **informa** business

First edition published 2020
by CRC Press
6000 Broken Sound Parkway NW, Suite 300, Boca Raton, FL 33487-2742

and by CRC Press
2 Park Square, Milton Park, Abingdon, Oxon, OX14 4RN

© 2021 Taylor & Francis Group, LLC

CRC Press is an imprint of Taylor & Francis Group, LLC

ISBN: 978-0-367-34383-5 (hbk)
ISBN: 978-0-367-49493-3 (pbk)
ISBN: 978-0-429-32545-8 (ebk)

Typeset in Times
by Lumina Datamatics Limited

# Contents

Foreword ........................................................................................ xiii
Preface .......................................................................................... xv
Acknowledgment ......................................................................... xvii
Authors ........................................................................................ xix

**Chapter 1**    Fundamentals of Energy and Energy Scenario .................... 1

    1.1    Fundamentals of Energy ................................................. 1
    1.2    Various Types of Energy ................................................. 1
        1.2.1    Potential Energy ................................................. 1
            1.2.1.1    Chemical Energy ................................. 1
            1.2.1.2    Nuclear Energy ................................... 1
            1.2.1.3    Stored Mechanical Energy ................. 2
        1.2.2    Kinetic Energy ................................................... 2
    1.3    Commercial and Non-commercial Energy ...................... 2
        1.3.1    Commercial Energy ............................................ 2
        1.3.2    Non-commercial Energy ..................................... 2
    1.4    Grades of Energy ............................................................ 3
        1.4.1    High-Grade Energy ............................................ 3
        1.4.2    Low-Grade Energy ............................................ 3
    1.5    Energy Demand and Supply in India ............................. 3
        1.5.1    Energy Demand ................................................. 3
        1.5.2    Energy Supply ................................................... 4
    1.6    Energy Scenario in India ............................................... 4
        1.6.1    Non-Renewable Energy Supply Options ........... 5
            1.6.1.1    Coal ..................................................... 7
            1.6.1.2    Petroleum/Oil ..................................... 8
            1.6.1.3    Natural Gas ......................................... 9
            1.6.1.4    Nuclear Energy ................................... 9
    1.7    Energy Security in India .............................................. 10
    1.8    Future Energy Strategy ................................................ 11
        1.8.1    Immediate Term Action Plan ........................... 11
        1.8.2    Medium Term Action Plan ............................... 11
        1.8.3    Long-Term Action Plan .................................... 11
    Questions ........................................................................... 12

**Chapter 2**    Energy Management and Energy Conservation Opportunities ......... 13

2.1    Energy Management ................................................................. 13
    2.1.1    Energy Management Techniques .................................. 13
        2.1.1.1    Self-Information and Attentiveness
                   Amongst the Crowds ..................................... 13
        2.1.1.2    Reengineering and Estimates........................ 14
        2.1.1.3    Updates of Technology................................. 14
2.2    Tariff ..................................................................................... 14
    2.2.1    Objectives of Tariff...................................................... 14
    2.2.2    Characteristics of a Tariff............................................. 15
    2.2.3    Various Kinds of Tariff ................................................ 15
2.3    Energy Conservation............................................................... 18
2.4    Energy Conservation Opportunities (ECOs) ........................... 19
    2.4.1    Energy Conservation in Households ............................ 19
    2.4.2    Energy Conservation in the Transport Sector .............. 23
    2.4.3    Energy Conservation in the Agriculture Sector ........... 24
    2.4.4    Energy Conservation for the Industries Sector............. 24
    2.4.5    Energy Conservation in Lighting ................................. 34
    2.4.6    Energy-Saving Opportunities in HVAC ...................... 36
Questions........................................................................................ 37

**Chapter 3**    Energy Audit ............................................................................ 39

3.1    Concept of Energy Audit........................................................ 39
3.2    Type of Energy Audit............................................................. 40
    3.2.1    The Type of Energy Audit to be Performed
             Depends On ................................................................ 40
    3.2.2    Preliminary Energy Audit Methodology....................... 40
    3.2.3    Detailed Energy Audit Methodology ........................... 41
        3.2.3.1    General Process for Detailed Audit
                   of Energy ..................................................... 41
        3.2.3.2    A Guide to Energy Audits Conducting
                   at a Glance ................................................... 43
3.3    Collecting Data Strategy......................................................... 46
3.4    Technical and Economic Feasibility ....................................... 47
3.5    Types of Energy Conservation Measures................................ 48
3.6    Understanding Energy Costs................................................... 48
    3.6.1    Fuel Costs ................................................................... 49
    3.6.2    Power Costs ................................................................. 49
3.7    Benchmarking and Energy Performance.................................. 49
3.8    Plant Energy Performance ...................................................... 51
    3.8.1    Production Factor ......................................................... 51
    3.8.2    Reference Year Equivalent Energy Use........................ 51
    3.8.3    Monthly Energy Performance ...................................... 52
3.9    Fuel and Energy Substitution ................................................. 52

3.10  The Energy Conservation Act, 2001 and Its Features............ 52
      3.10.1   Policy Framework—Energy Conservation
               Act—2001................................................................52
      3.10.2   Important Characteristics of the Energy
               Conservation Act....................................................53
      3.10.3   Designated Consumers............................................53
      3.10.4   Certification of Energy Managers
               and Accreditation of Energy Auditing Firms.......... 53
      3.10.5   Energy Conservation Building Codes ....................54
      3.10.6   Central Energy Conservation Fund........................54
      3.10.7   Bureau of Energy Efficiency (BEE) ......................54
               3.10.7.1   Role of Bureau of Energy Efficiency ..... 55
               3.10.7.2   Role of Central and State Governments ... 55
               3.10.7.3   Enforcement Through Self-Regulation .... 55
               3.10.7.4   Penalties and Adjudication ....................55
3.11  Responsibilities and Duties to be Assigned Under the
      Energy Conservation Act, 2001................................................56
      3.11.1   Energy Manager: Responsibilities...........................56
      3.11.2   Energy Manager: Duties..........................................56
      3.11.3   Energy Auditors: Responsibilities ..........................57
      3.11.4   Energy Auditors: Duties .........................................57
3.12  Energy Audit Instruments .........................................................57
      3.12.1   Electrical Measuring Instruments ...........................57
      3.12.2   Energy Meter..........................................................58
      3.12.3   Combustion Analyzer..............................................58
      3.12.4   Fuel Efficiency Monitor..........................................58
      3.12.5   Fyrite .....................................................................58
      3.12.6   Contact Thermometer..............................................58
      3.12.7   Infrared Thermometer.............................................59
      3.12.8   Pitot Tube and Manometer .....................................59
      3.12.9   Water Flow Meter...................................................59
      3.12.10  Speed Measurements..............................................59
      3.12.11  Leak Detectors ......................................................59
      3.12.12  Pyranometer ..........................................................59
      3.12.13  Lux Meters.............................................................60
      3.12.14  Anemometer ..........................................................60
Questions................................................................................................60


Chapter 4   Material and Energy Balance................................................................63

4.1   Basic Principles of Energy and Mass Balances......................63
4.2   Usage and the Working of the Sankey Diagram ....................65
4.3   Material Balances....................................................................65
      4.3.1   Basis and Units .......................................................66
      4.3.2   Net Mass and Composition......................................66

4.3.3    Concentrations .......................................................... 67
4.3.4    Types of Process Situations ..................................... 69
     4.3.4.1    Continuous Processes ..................................... 69
     4.3.4.2    Blending ......................................................... 70
     4.3.4.3    Drying ........................................................... 70
4.4    Energy Balances ...................................................................... 71
    4.4.1    Heat Balances ........................................................ 71
    4.4.2    Entry of Heat ......................................................... 74
    4.4.3    Left Out Heat ........................................................ 74
    4.4.4    Types of Energy .................................................... 74
    4.4.5    Summary ................................................................ 75
4.5    Preparation of Process Flow Chart ....................................... 76
4.6    Facility as an Energy System ................................................. 77
    4.6.1    Boiler System ........................................................ 78
    4.6.2    Cooling Tower ...................................................... 78
    4.6.3    System of Compressed Air ................................... 79
4.7    Procedures in Carrying Out Balancing in Energy and Mass ...... 79
    4.7.1    Guidelines for Material and Energy Balance ............... 79
    4.7.2    Procedure for Energy and Mass Balance Calculation ..... 80
Questions ............................................................................................ 81

**Chapter 5**    Energy Action Planning ........................................................ 83

5.1    Energy Management System ................................................... 83
    5.1.1    Support and Commitment of the Top Management ...... 83
     5.1.1.1    Energy Manager Appointment ...................... 83
     5.1.1.2    Dedicated Energy Team Formation .............. 84
     5.1.1.3    Establishment of Energy Policy .................... 84
    5.1.2    Performance of Energy Assessment ............................ 85
     5.1.2.1    Collection of Data and Management ............. 85
     5.1.2.2    Baseline Establishment ................................. 87
     5.1.2.3    Benchmark ..................................................... 87
     5.1.2.4    Analysis and Evaluation ............................... 88
     5.1.2.5    Conduct Technical Assessments & Audits ..... 89
    5.1.3    Goals Setting in an Energy Management
           Organization ........................................................... 89
    5.1.4    Formulation of an Action Plan .................................. 90
    5.1.5    Implement and Execution of Action Plan ................... 92
    5.1.6    Process of Evaluating Progress ................................. 93
    5.1.7    Achievement Recognition: A Tool for Motivation ........ 93
Questions ............................................................................................ 94

**Chapter 6**  Monitoring and Targeting ................................................. 95

    6.1   Introduction .............................................................. 95
          6.1.1   A Case Study: Toray Textiles (EUROPE) Ltd. ............ 96
    6.2   Elements of Targeting & Monitoring System ........................... 97
    6.3   Analysis of Data and Information ............................................ 97
    6.4   Application ............................................................................. 98
          6.4.1   Energy and Production ................................................ 98
          6.4.2   Development of Energy Performance Model .............. 98
                 6.4.2.1   Step 1: Plot Energy Usage vs. Production ...... 99
                 6.4.2.2   Step 2: Determine the Baseline
                                 Relationship ..................................... 99
    6.5   Cumulative Sum Control Chart ............................................ 102
          6.5.1   Step 1: Calculate the CUSUM ................................. 102
          6.5.2   Step 2: Interpret the CUSUM Graph ........................ 103
          6.5.3   Summary: Regression and CUSUM .......................... 103
    Questions ...................................................................................... 103

**Chapter 7**  Electrical Energy Management ...................................... 105

    7.1   Maximum Demand Basics ..................................................... 105
    7.2   Improvement of Power Factor ............................................... 106
    7.3   Capacitor Locations ............................................................. 106
    7.4   Pump .................................................................................... 107
          7.4.1   Classification of Pumps ........................................... 107
                 7.4.1.1   Positive Displacement Pumps ...................... 107
                 7.4.1.2   Rotodynamic Pumps ..................................... 108
                 7.4.1.3   Other Types of Pumps ................................... 109
    7.5   Fan Systems ......................................................................... 111
          7.5.1   Types of Fan ............................................................ 111
                 7.5.1.1   Types of Centrifugal Fan .............................. 112
                 7.5.1.2   Types of Axial Flow Fans ............................. 113
          7.5.2   Assessment of Performance of Fans .......................... 113
                 7.5.2.1   Measurement of the Flow of Air .................. 113
                 7.5.2.2   Velocity Pressure/Velocity Calculation ........ 114
                 7.5.2.3   Efficiency of Fan ......................................... 115
          7.5.3   Energy Conservation Opportunities in Fans .............. 115
    7.6   Energy-Efficient Motors ....................................................... 116
          7.6.1   Salient Features of Motor Performance ..................... 117
          7.6.2   Stator and Rotor $I^2R$ Losses ..................................... 118
          7.6.3   Core Losses ............................................................. 119
          7.6.4   Friction and Windage Losses .................................... 119

7.6.5   Stray Load-Losses ....................................................... 119
7.6.6   Technicality Related to the Motors for Energy
        Efficiency.................................................................. 120
Questions .................................................................................. 120

**Chapter 8**   Thermal Energy Management........................................................... 123

8.1   Boiler...................................................................................... 123
      8.1.1   Boiler Types and Classification ................................... 124
      8.1.2   Boiler Performance Evaluation ................................... 124
      8.1.3   Energy Conservation Opportunities in Boiler............ 125
8.2   Industrial Heating System.................................................. 128
      8.2.1   Classification of Furnaces .......................................... 128
              8.2.1.1   Oil Fired Furnace ........................................ 128
              8.2.1.2   Typical Furnace System ............................... 128
              8.2.1.3   Rerolling Mill Furnace................................. 128
      8.2.2   Furnace Heat Transfer ................................................ 130
              8.2.2.1   Types of Continuous Reheating Furnace ..... 131
      8.2.3   Furnace Energy Supply ............................................... 133
      8.2.4   Furnace Performance Evaluation ................................ 133
8.3   Fluidized Bed Combustion (FBC) Boilers............................ 134
      8.3.1   Introduction to FBC Boilers ....................................... 134
      8.3.2   Mechanism of Fluidized Bed Combustion.................. 134
8.4   Cogeneration ....................................................................... 135
      8.4.1   Need for Cogeneration................................................ 135
      8.4.2   Steam Turbine Cogeneration Systems........................ 135
      8.4.3   Gas Turbine Cogeneration Systems............................ 136
      8.4.4   Reciprocating Engine Cogeneration System .............. 136
      8.4.5   Classification of Cogeneration Systems ..................... 136
8.5   Biomass Utilization in FBC and Co-generation Technology ...... 137
8.6   Waste Heat Recovery .......................................................... 138
      8.6.1   Benefits of Waste Heat Recovery ............................... 138
              8.6.1.1   Direct Benefits............................................. 138
              8.6.1.2   Indirect Benefits .......................................... 138
      8.6.2   Development of a Waste Heat Recovery System......... 139
              8.6.2.1   Commercial Waste Heat Recovery
                        Devices......................................................... 139
8.7   Thermal Insulation.............................................................. 140
Questions .................................................................................. 141

**Chapter 9**   Building Energy Management .......................................................... 143

9.1   Introduction......................................................................... 143
9.2   Factors Affecting Climate................................................... 144

9.3 Energy Conservation Building Code-Provisions in the EC Act 2001 ............................................................. 149
   9.3.1 E.C.B.C. Development ............................................. 149
   9.3.2 Broad Stakeholder Participation ............................. 150
   9.3.3 Features of E.C.B.C. ................................................ 150
   9.3.4 E.C.B.C.—Major Elements of the Building Plan .... 151
   9.3.5 E.C.B.C. Benefits .................................................... 151
   9.3.6 E.C.B.C. Implementation—All India Level ............ 152
   9.3.7 Road Map to Make E.C.B.C. Mandatory ............... 152
9.4 Energy Conservation Measures ........................................... 152
   9.4.1 Building Envelope .................................................. 152
   9.4.2 Ventilation and Indoor Air Quality ....................... 152
   9.4.3 Electrical Systems .................................................. 153
9.5 Commercial and Industrial Buildings ................................. 156
   9.5.1 Building-Envelope Technologies ........................... 156
      9.5.1.1 Light-Pipe Technologies ......................... 156
      9.5.1.2 HVAC Systems and Controls .................. 156
      9.5.1.3 Cogeneration ........................................... 157
9.6 Energy Saving in Buildings Due to Various Passive Systems ............................................................................... 157
9.7 Barriers in Adopting Energy Efficiency in Residential Buildings ............................................................................. 157
Questions ..................................................................................... 158

Chapter 10 Economic Analysis and Project Planning Techniques ..................... 159
10.1 Life Cycle Cost (LCC) and Life Cycle Assessment (LCA) Method .................................................................... 159
   10.1.1 Life Cycle Assessment (LCA) Method .................. 159
10.2 Levelized Cost of Energy .................................................... 160
10.3 Simple Payback Period ........................................................ 161
10.4 Time Value of Money ........................................................... 162
10.5 Return on Investment (ROI) ................................................ 163
10.6 Net Present Value (NPV) ..................................................... 163
10.7 Internal Rate of Return ........................................................ 165
10.8 Cash Flows .......................................................................... 167
10.9 Sensitivity Analysis ............................................................. 168
10.10 Project Planning Techniques; CPM and Pert ...................... 168
Questions ..................................................................................... 171

References ............................................................................................ 173
Index .................................................................................................... 175

# Foreword

Energy plays an important role in the economic development of any country. The energy sector assumes a critical importance in view of the ever-increasing energy needs which requires huge investments. The per capita consumption of energy in India is low when compared to many countries. Energy management is necessary for production, procurement, and proper utilization of energy, which ultimately minimizes energy costs and waste without affecting the production and quality.

Energy conservation through increasing energy efficiency is the best solution which can be done through proper energy management and audit. Energy audit is the key to a systematic approach for decision making in the area of energy management and it provides a unique pathway for customers to save money. Energy can be saved through conservation or increasing the efficiency of the system. The energy management and auditing methods are also essential during the management of natural hazards.

It gives me immense pleasure to record that the authors of the book "Energy Management: Conservation and Audits" have made very sincere efforts in encompassing the different aspects of energy management, conservation, and audit. The authors have a strong background in the field of energy engineering. The positive impact of energy conservation practices on the reduction of energy cost and establishment of maintenance parts have been elaborated in a very lucid manner for the readers. This book would be very helpful for students and professionals working in the field of energy management and audit. The authors deserve appreciation for writing this valuable book.

**Dr. Mrutyunjay Mohapatra, Ph.D.**
*Director General, India Meteorological Department*
*Permanent Representative of India with WMO*
*Member of Executive Council, WMO*

Reducing or even eliminating the greenhouse gas emissions and local pollution associated with energy use in building is a global challenge. Whilst the goal of achieving process efficiency in industry or occupant comfort and amenity in buildings with the least use of fossil fuels is easily understood, the most viable means of satisfying this goal are not. This is because a sea of options are present while navigating these in the purpose of this book. It systematically details how to understand and measure the energy characteristics of processes and buildings. A comprehensive overview of relevant technologies is also provided. Building on this, the book provides a clear blueprint on how this knowledge should be best used with a chapter dedicated to "Energy Action Planning."

This book will be ideally useful to educators and students. It is a comprehensive resource for the practicing Energy Managers as they make critical decisions whose cumulative input will have a profound influence on our planet's future.

**Professor Brain Norton, D.Sc., MRIA**
*Principal, Technological University Dublin*

# Preface

Energy resources present on the earth are in diminishing nature. These are depleting fast with time as the uses are increasing exponentially. Though, most of the resources present in earth atmosphere are in renewable nature. However, many economically viable technologies are already developed to harness renewable energy resources. Only 5% of total power is generated from available technologies. The capital investments for the development of technology are very high as compared to use them effectively. The main idea behind energy management is to control the use of these resources or increases the life of diminishing energy resources. This will come up with lesser goods and services cost which will be causing the least environmental effect. This book deals with the concept of energy management through conservation and audits.

Chapter 1 deals with the concept of energy and various types of energy along with different grades of energy. Energy demand and supply, energy scenario and energy security in India and future energy action plan/strategies are also discussed.

Chapter 2 discusses about energy management, energy management techniques, and energy conservation opportunities. The aim of energy management is to maintain and accomplish the maximum energy obtaining and using throughout the organization. It reduces the cost of energy/misuse without affecting quality, and production without affecting environment. Energy tariff objectives, characteristics and their different types are also discussed in context of their merits and demerits.

Chapter 3 includes the positive impact of energy audit on the reduction of energy cost and establishment of maintenance parts. It also controls the quality of the products, which are directly related to the utility of the energy and the production. The primary aim of an energy audit is to find out various pathways for reducing the consumption of energy per unit of the output of the product or reducing the cost of operation. The concept of energy audit with the practical approaches has been analyzed and discussed in this chapter.

Chapter 4 deals with the material and energy balance. The balancing is an essential aspect of energy and materials in industrial processes. The material and energy balance can affect the production cost. Therefore, it is very essential to do material balance for reducing the consumption of raw material or reducing the waste formation during the process. By selecting the proper operation process, the energy to be used during that process operation can be optimized which can reduce the electricity bill finally by production cost with the help of energy balance. The main reason for carrying material and energy balance is for identification of the unknown process taking place. The prime effect and process of material and energy balance is also discussed in this chapter.

Chapter 5 deals with the purpose of goals setting in an energy management organization, formulation of action plan, implement-cum-execution of action plan, evaluation of progress in an organization, and the importance of recognition of achievement. Without goals setting, no organization can fulfill its purpose, while action plan is necessary for the manifestation of the blueprint of the energy project on the ground.

How evaluation of progress opens new pathways for energy efficiency and suggests corrective measures for the energy loss and destructive operations and actions are discussed in this chapter. This chapter also throws light on how recognition dissolve in the blood of an individual or of a team as a whole in the form of motivation, which leads an employee to put all his efforts to achieve organizational goals.

Chapter 6 discusses the difference between monitoring and targeting with the effects of main elements of system and its benefits. Targeting and monitoring are one of the management techniques where essential facilities for production are taken as a controlling factor such as the raw materials, energy and other goods. It divides the facilities systematically into divisions related to energy cost. With the help of such data, target and planning can be done to find out the anomaly in the data, actions for remedy, and finally implementation.

Chapter 7 discusses electrical energy management based on improvement of power factor its usefulness, energy conservation opportunities, differences between conventional motor, and energy efficient motor, etc. This chapter also deals with different types of motors, pumps, fans and their losses and energy conservation opportunities.

Chapter 8 explains the thermal energy generation and management in the boilers. It deals with the proper utilization of the resources and the way to protect it or preserve it for the longer time for the future generations. The various types of heating systems, their operations, and various heat recovery systems are being analyzed. Therefore, the proper usage of thermal energy has also been understood and the minimum losses that take place during the operation is studied.

Chapter 9 explains the possibilities of energy conservation in buildings such as commercial offices and domestic sector which devour around 33% of the aggregate power utilized in the nation. Due to various difficulties in having energy through renewable, the conservation through green buildings cannot be ignored. In this chapter, various parameters have been discussed—related to human comfort, proper utilization of daylight and thermal insulation. The energy conservation building codes development and its effects on energy saving opportunities in building are also discussed in detail.

Chapter 10 deals with the interrelated nature of economic growth, natural resource use, and environmental protection. Economic development, the ultimate goal of which is to improve human welfare, is crucially dependent on the environment and natural resources to provide the goods and services which directly and indirectly generate socioeconomic benefits. This chapter also discusses about the economic analysis, scope and goal definition for life cycle assessment, and its effect on project planning techniques.

I hope this book is complete in all aspects of energy management. It would serve as a useful tool for learners, faculty members, practicing engineers, and students. Despite our best efforts, we regret if some errors are in the manuscript due to inadvertent mistake. We will greatly appreciate being informed about errors and receiving constructive criticism for further improvement of the book.

**Anil Kumar**
**Om Prakash**
**Prashant Singh Chauhan**
**Samsher**

# Acknowledgment

This book is a tribute to the engineers and scientists who continue to push forward the practices and technologies of energy conservation and management. This book work would not be completed without the efforts of numerous individuals including the primary writers, contributing authors, technical reviewers, and practitioners. Our first and foremost gratitude goes to God Almighty for giving us the opportunity and strength to do our part of service to the society.

We express our heartfelt gratitude to Prof. Yogesh Singh, Hon'ble Vice Chancellor, Delhi Technological University, Delhi (India); Hon'ble Vice Chancellor, Birla Institute of Technology, Mesra, Ranchi, India and Ms. Harjot Kaur Bamhrah (IAS) Principal Secretary- Department of Science and Technology, Government of Bihar for their kind encouragement.

We would like to express our thanks to many people, faculty, and friends who provided valuable input during preparation of this text. We thank especially Prof. G. N. Tiwari, Centre for Energy Studies, Indian Institute of Technology Delhi, India; Prof. Perapong Tekasakul, Vice President, Research System and Graduate Studies, Prince of Songkla University, Hat Yai, Songkhla, Thailand; and Prof. Emran Khan, Head, Department of Mechanical Engineering, Jamia Millia Islamia, New Delhi for their valuable suggestions and help.

We wish to acknowledge Prof. Vipin, Head, Mechanical Engineering Department, Delhi Technological University, Delhi, India; Prof. Anil Kumar, Former Principal, Gaya College of Engineering, Gaya, Bihar; Prof. Nirmal Kumar, Principal, Gaya College of Engineering, Gaya, Bihar; Prof. Amit Pal, DTU-Delhi; Dr. Girish Kumar, DTU-Delhi; Prof. Prashant Baredar, Head, Energy Centre, Maulana Azad National Institute of Technology, Bhopal, India and all colleagues for their support and encouragement. We also appreciate the support of our scholars Sanjeev Kumar Bhukesh, Pranshu Shrivastava, Sankalp Kumar and Aviral Agrawal.

We appreciate our spouses, Mrs. Abhilasha, Mrs. Poonam Pandey, Mrs. Preetu Singh Chauhan, Mrs. Sunita Gautam and our beloved children Master Tijil Kumar, Ms. Idika Kumar, Ms. Shravani Pandey, Master Abhinandan Singh Chauhan, Master Abhineet Singh Chauhan, Dr. Pratibha Gautam, and Mr. Manish Gautam. They have been a great cause of support and inspiration, and their endurance and sympathy throughout this project have been most valued. Our heartfelt special thanks go toward CRC Press, for publishing this book. We would also like to thank those who directly or indirectly were involved in bringing up this book successfully.

Finally, yet importantly, we wish to express our warmest gratitude to our respected parents Late Sh. Tara Chand, Smt. Vimlesh, Sh. Krishna Nandan Pandey, Smt. Indu Devi, Sh. Nagendra Singh Chauhan, Smt. Binita Singh Chauhan, Late Sh. Param Lal Gautam, Smt. Radha Gautam and our siblings for their unselfish efforts to help in all fields of life.

**Anil Kumar**
**Om Prakash**
**Prashant Singh Chauhan**
**Samsher**

# Authors

**Dr. Anil Kumar** is associate professor in the department of mechanical engineering, Delhi Technological University, Delhi, India. He has completed his PhD in solar energy from Centre for Energy Studies, Indian Institute of Technology Delhi, India in the year 2007. He was a postdoctoral researcher at Energy Technology Research Center, department of mechanical engineering, faculty of engineering, Prince of Songkla University, Hat Yai, Songkhla, Thailand in the discipline of energy technology from June 2015 to May 2017. He has also served as assistant professor at Energy Centre, Maulana Azad National Institute of Technology Bhopal, India from 2010 to 2018, and assistant professor in the department of mechanical engineering, University Institute of Technology, Rajiv Gandhi Proudyogiki Vishwavidyalaya, Bhopal, India from 2005 to 2010. His nature of experience is teaching and research (science, technology, society, and sustainable development). His areas of specializations are; energy technology, renewable energy, solar energy applications, energy economics, heat transfer, natural rubber sheet drying, and environmental issues. He has successfully completed many research funded projects in these areas. He has more than 14 years of experience in the field of energy technology. He has published 115 papers in international peer-reviewed journals and 75 papers in the international/national conferences proceeding. His paper is being cited in all the reputed relevant journals. He has received more than 2700 citations with 28 h-index. He has developed thin-layer drying model with his student in 2014. It is known as "Prakash and Kumar model." This model is being used and cited by many researchers around the globe.

He has also published eight books with international/national publishers. He has filed and published three patents—one in the solar dryer and another in the solar photovoltaic thermal collector. He has supervised 7 PhD scholars and 32 master students. Dr. Kumar has visited countries namely UK, Thailand, and Malaysia. He has received various awards and appreciation from reputed sources. Some of them are presented below:

- **Commendable Research Award 2019**: For publishing quality research paper in Delhi Technological University, Delhi, India.
- **Research Excellence Award 2016**: The researcher has Top 20 Publications from the Web of Science database, honored by Dr. Chusak Limsakul, President Prince of Songkla University, Hat Yai, Thailand.

- Appreciation for outstanding service in the reviewers' committee and active reviewers during the year 2015–2017 from International Institute of Engineers, Kolkata and Elsevier.
- Best Paper Award in Global Conference on Energy and Sustainable Development (GCESD 2015), February 24–26, 2015 at Coventry University Technology Park, Puma Way, Coventry, West Midlands CV1 2TT.

**Dr. Om Prakash** is working as assistant professor in the department of mechanical engineering, Birla Institute of Technology, Mesra, Ranchi, India. This university is the premier leading technical university in India, which is established in 1955 by Mr. B. M. Birla. Total number of registered students across all centers—more than 10,000. Doctoral students—150. Programs offered—Undergraduate, Postgraduate, and Doctoral. He has published about 60 research papers in international/national journals, international/national conferences and book chapters. He has also published two patents. He has published three edited books. He has developed one unique mathematical model namely "Prakash and Kumar model." He has one government funded project of 18.5 lakhs. He is the reviewer in many reputed international journals. He is working in the field of renewable energy, energy and environment, renewable energy, solar energy applications, heat transfer, and energy economics.

**Dr. Prashant Singh Chauhan** is head of the department of mechanical engineering, Gaya College of Engineering, Gaya, Bihar, India. He has completed his PhD (2017) and MTech (2011) from department of energy (Energy Centre), Maulana Azad National Institute of Technology, Bhopal, India. He was a postdoctoral researcher in the department of mechanical engineering, faculty of engineering, Prince of Songkla University, Hat Yai, Thailand. He has also served as assistant professor in the department of mechanical engineering, People's College of Research and Technology, People's University, Bhopal, India from 2011 to 2013. His nature of experience is teaching and research in the field of energy, technology and sustainable development. His areas of specializations are renewable energy, solar energy applications, energy technology, heat transfer, and environment. He has completed many student research projects in these areas. He has more than 7 years of experience in the field of energy. He has published 10 papers in international peer-reviewed journals, 6 papers in the international/national conferences proceeding and one book chapter. His paper is being cited in all the reputed relevant journals. He has received more than 160 citations and the h index by Google Scholar is 6. He has one ongoing research project funded by AICTE and NPIU-MHRD of Rs. 20.92 Lakhs. He has supervised 15 BTech students. He is a reviewer of many reputed international journals (Elsevier, Taylor Francis, and Springer, etc.).

 **Dr. Samsher** has completed his PhD from the department of mechanical engineering, Indian Institute of Technology Delhi, India in 2005. At present, he is a professor and has additional charge of Registrar and Controller of Finance in Delhi Technological University, Delhi, India. He is having a vast experience of 33 years in academics, research, and industries. He has also worked as a director, National Institute of Electronics and Information Technology, (Ministry of Electronics and Information Technology, GOI), Delhi, professor in National Institute of Technology, Jalandhar, India, and deputy director in National Power Training Institute, Faridabad, India and Engineer in National Thermal Power Corporation (NTPC) India. He has also been a member of Academic Council and Board of Management, Delhi Technological University, Delhi. His areas of specializations are thermal engineering, power plant engineering, energy technology, and renewable energy. He has published more than 70 papers in international peer-reviewed journals and proceedings. He has supervised 5 PhD scholars, 17 masters' students. Prof. Samsher has visited countries namely USA, UK, Mexico, Singapore, and Malaysia for academic and research purposes. He has received various awards and appreciation from reputed sources, such as, "Commendable Research Award" for publishing quality research paper in DTU, Delhi in 2018, "Consistency High Performance Awards" in NPTI in 2000, "Scroll of Honour" in NPTI in 2000, and "Subject Prize" by Institution of Engineers (India) in 2011.

# 1 Fundamentals of Energy and Energy Scenario

## 1.1 FUNDAMENTALS OF ENERGY

Capacity to do work is said to be energy, whereas work is defined as the transfer of energy from one system to another system. It can be practically used to change the systems around us, for example, moving body muscles, using electricity, using mechanical devices. It is of different types, namely, heat (thermal), mechanical, light (radiant), chemical, electrical, and nuclear energy.

## 1.2 VARIOUS TYPES OF ENERGY

Energy is of two types, namely, stored energy or potential energy and kinetic energy. For example, chemical energy contained in food is stored in our body until it is released for work.

### 1.2.1 POTENTIAL ENERGY

It is the energy stored in a substance. The energy has the potential to do work. Gravity gives potential energy to a subject. This potential energy is a result of gravity pulling downwards. It occurs in various forms.

#### 1.2.1.1 Chemical Energy

It is the energy, which is stored in chemical bonds of any chemical compounds. It is released during a chemical reaction in the form of heat and as a byproduct. The process is called exothermic reaction. Examples of stored chemical energy are batteries, petroleum, biomass, natural gas, and coal. Once the chemical energy is released from a substance, it is then transformed into an entirely new material. Like, during an explosion, chemical energy, which has been stored in it, transferred to the surroundings in the form of thermal energy, sound energy, and kinetic energy.

#### 1.2.1.2 Nuclear Energy

Nuclear energy is the energy found in the nucleus of an atom. Atoms are those tiny units that make up matter in the universe. Nucleus is held together by energy. Electricity can be produced by using this energy. It can be obtained in two ways—nuclear fusion and nuclear fission. In nuclear fusion, the energy is released when the atoms are combined to form a larger atom. In nuclear fission, the atoms are split into smaller atoms, releasing energy.

### 1.2.1.3   Stored Mechanical Energy

Energy that has been stored in objects by the application of a force is called stored mechanical energy.

### 1.2.2   KINETIC ENERGY

All the moving objects have kinetic energy in them. It is the energy, which is possessed by any object due to motion. It is of various types.

1. **Radiant Energy**: Radiation is electromagnetic energy, which includes visible light, X-rays, gamma rays, and radio waves.
2. **Thermal Energy**: Demand of conventional energy sources can be reduced by the utilization of thermal energy storage (TES) for a couple of reasons. Firstly, they can help create a balance between the supply of energy and the demand in power by generating electricity from renewable energy sources. Secondly, the final energy consumption can be reduced by the utilization of waste heat in industrial process.
3. **Sound**: It is the movement of energy through substances in longitudinal (compression/rarefaction) waves.
4. **Electrical Energy**: Electrical energy is the energy carried by the movement of electrons in an electrical conductor. It is relatively easy to transmit and use, and thus it is a highly useful form of energy. It is generated, when the electrons are allowed to move on a particular path in any conducting substance like a wire.

## 1.3   COMMERCIAL AND NON-COMMERCIAL ENERGY

### 1.3.1   COMMERCIAL ENERGY

These are types of energy, which can be usually found in the market at a fixed rate. Significant types of such energy are known as electricity, coal, and refined petroleum products. Commercial fuels are a predominant source of economic growth as well as the general population also use it for many household tasks.

### 1.3.2   NON-COMMERCIAL ENERGY

Non-commercial energy is a type of energy, which cannot be found in the commercial market. It is used in rural areas for domestic purposes like cooking, drying, heating of water, etc. Sources of non-commercial are firewood, cattle dung cake, municipal waste, and agricultural waste. They are known as non-conventional fuels, and most of the times are ignored during energy accounting.

## 1.4 GRADES OF ENERGY

### 1.4.1 HIGH-GRADE ENERGY

Electrical and chemical energy are known as high-grade energy since their energy density is very high. Small amount of energy can produce a considerable amount of work. The molecules are highly ordered and compact that store these forms of energy and therefore, are considered to be high-grade energy. These are better to use for high-grade applications such as melting metals than simple heating of water.

### 1.4.2 LOW-GRADE ENERGY

Heat is a prime example of low-grade energy. The molecules, which store this type of energy (solid and liquid molecules), are randomly distributed. This disordered state of the molecules and the dissipated energy are categorized as low-grade energy.

## 1.5 ENERGY DEMAND AND SUPPLY IN INDIA

### 1.5.1 ENERGY DEMAND

It has been projected that the worldwide consumption of oil, natural gas, and other energy sources would increase by more than 40% in the year 2035.

This increase in demand will be due to the rise in population, which has been predicted to increase by 25% in the next 20 years especially in countries with emerging economies, such as China and India where improved standards of living, and the rising energy demand from economic output is more likely to add pressure on energy supplies.

According to the International Energy Agency (IEA), output generated from resources such as oil sands and other heavy oils has increased massively by ten-fold since the year 1980 and not surprisingly is set to rise by quadruple by 2035. The rapid development of liquid-rich shale's, which uses the same hydraulic fracturing technologies like shale gas, holds much promise if it can be extended globally.

As per WEO New Policies Scenario (NPS) plan, it has been announced by the countries to face the security of energy, change in climate and various challenges related to energy. The 450 scenarios of WEO set a course of energy to have a 50% chance of achieving goals of curbing the rise in average global temperature to about 2°C when a comparison is made with the pre-industrial levels.

Economic structure is reflected by the demand of sectoral energy of a particular country. In the year 2017, India's largest energy consumer was the building sector and represented 218 Mtoe of the total country demand for the primary energy, which had biomass as the dominant fuel. Industrial sector demand will remain consistent till 2035.

India accounts for more than a quarter of net global primary energy demand growth between 2017 and 2040. 42% of this new energy demand is met through coal. It means $CO_2$ emissions will roughly double by 2040. Gas production grows but fails to keep pace with demand, implying a significant growth in gas imports.

This has been attributed to the rising demand for electricity in industries and residential/commercial activities. Having a similar direction, the share of power sector has been expectedly rising to about 42% by 2035 under NPS.

### 1.5.2 ENERGY SUPPLY

In India, the domestic energy production has grown from 502 Mtoe in the year 2009 to 775 Mtoe in 2013 with a compound annual growth rate of 2.9%. Assuming demand growth at a compound annual growth rate of 4% during the same period, domestic supply was not able to keep up its pace with the demand. Largest production source was biomass having a 46% share in 1990; however, it got reduced to 33% in 2009. Coal had the highest production volume increment as raised from 244 Mtoe in 2009 to 341 Mtoe in 2013 at a compound annual growth rate of 4.6%.

Coal represents about one half of the energy production domestically. Natural gas, however, is the fastest-growing fuel with a rise in energy production of domestic energy. The second-largest source was hydro energy, as it accounted for 20% of installed capacity.

## 1.6   ENERGY SCENARIO IN INDIA

Primary energy supply which has been normalized concerning the GDP and population for the year 2017. Table 1.1 gives the comparison of India with other regions of the world:

Table 1.1 clearly depicts that per capita electricity consumption in India is very low as compared to the world. It utilized 947 kWh in 2017 compared to 3152 kWh by

### TABLE 1.1
### World Energy Statistics

| Countries/Regions | Population (millions) | GDP Per Capita (PPP) 2010 USD | TPES/Pop toe/Capita (toe) | TPES/GDP (toe-2010 USD) | Elec. Cons/Pop (kWh) | $CO_2$/ GDP- 2010 PPP |
|---|---|---|---|---|---|---|
| World | 7519 | 113,555 | 1.86 | 0.17 | 3152 | 0.29 |
| OECD | 1295 | 50,410 | 4.10 | 0.10 | 7992 | 0.23 |
| Middle East | 237 | 5344 | 3.17 | 0.32 | 4132 | 0.33 |
| Non-OECD Asia | 2501 | 18,743 | 0.75 | 0.28 | 1085 | 0.22 |
| Non-OECD Europe & Eunasia | 340 | 5641 | 3.30 | 0.41 | 4581 | 0.44 |
| China | 1394 | 21,201 | 2.21 | 0.29 | 4555 | 0.44 |
| Non-OECD America | 497 | 6509 | 1.23 | 0.14 | 2060 | 0.16 |
| Africa | 1255 | 5708 | 0.65 | 0.33 | 574 | 0.21 |
| India | 1339.2 | 8436.9 | 0.66 | 0.34 | 947 | 0.26 |

*Source:*  Birol, F. *Key World Energy Statistics*, International Energy Agency, 2017.

rest of the world, 7992 kWh by the OECD countries, 4555 kWh by China. However, the efficiency of energy use for producing the gross domestic product for purchasing power parity of India is higher than many countries. Over the period 2021–22, it is projected that there will be a growth rate of 9% TPES requirement. Total 85% of villages have been electrified till now, and around 84 million households in the country still do not have any access to electricity.

It has been noticed that for utilization of over 4000 kWh/person, the graph plateaus out and eventually straighten up. Even those who have power connections have to suffer from power shortages. Consumers and the economy bear a large load due to the bad quality of power supply. As a backup for low voltages, unscheduled power cuts or variable frequency, industries maintain diesel-powered generators whereas the households have invertors with batteries as a backup for low voltages, unscheduled power cuts or variable frequency. Equipment are frequently damaged (due to the erratic electricity supply).

The cost of idle human resources should be added to the loss of production if the power supply has been interrupted. Total amount of electricity has increased by the rate of 7.2%/annum over the last 25 years. This shows an improvement in plant load factor (PLF). However, the consumption is still restricted as power shortages have continued to be a hindrance to the development of the country.

The shortages are also attributed to meager investments in transmission and distribution. The huge investment has been attracted by the increasing generation capacity. Consequently, the state's Electricity Board remains financially weak and is unable to generate finances for further investment. The extent of energy deficit varies from one state to another. Table 1.2 gives the comparative status of power supply in various states and union territories in India. It also provides an insight into the peak demand for power and the shortages in peak power.

The history of focusing on financing in power generation results in loading more power on an inadequate transmission and distribution (T&D) network. Power cannot be easily moved from areas of surplus to those in deficit as the T&D investments have not kept pace with investments in generation. Industrial and commercial organizations are forced to have standby energy generation to meet demand or to provide quality supply on a 24 × 7 basis to support critical processes and give peaking support.

## 1.6.1 Non-Renewable Energy Supply Options

Action plans to meet the energy demand of India has been restricted by the energy resources and import opportunities of the country. However, India is not very equipped with natural sources of energy. Oil, gas, and uranium reserves are very less. However, there is a large reserve for thorium. Even though coal is abundant, but it is regionally available and has a high ash content and low calorific value. Its advantage is that it has low sulfur content. Potential of hydro energy is enormous, but it is very small as compared to the needs of India, and energy content is very less also. There is also a need to analyze the environmental as well as the social impacts of storage. The higher gestation period of hydro projects also leads to huge costs.

**TABLE 1.2**
**India's Energy Supply and Demand**

| States/UT/Region | Power Supply (in MU) | | | Peak Demand (in MW) | | |
|---|---|---|---|---|---|---|
| | Requirement | Availability | Surplus/Deficit (%) | Requirement | Availability | Surplus/Deficit (%) |
| Chandigarh | 1413 | 1413 | 0 | 301 | 301 | 0 |
| Delhi | 23,863 | 23,800 | −0.3 | 4810 | 4739 | −1.5 |
| Haryana | 31,762 | 29,912 | −5.8 | 6142 | 5574 | −9.2 |
| Himachal Pradesh | 6964 | 6713 | −3.6 | 1278 | 1187 | −7.1 |
| Jammu & Kashmir | 12,427 | 9268 | −25.4 | 2500 | 1690 | −32.4 |
| Punjab | 41,226 | 38,649 | −6.3 | 9399 | 7938 | −15.5 |
| Rajasthan | 40,956 | 40,552 | −1.0 | 7582 | 7408 | −2.3 |
| Uttar Pradesh | 70,098 | 59,306 | −15.4 | 11,082 | 10,672 | −3.7 |
| Uttarakhand | 9022 | 8451 | −6.3 | 1520 | 1520 | 0 |
| Chhattisgarh | 9250 | 9096 | −1.7 | 2913 | 2759 | −5.3 |
| Gujarat | 65,217 | 61,199 | -6.2 | 10,786 | 9947 | −7.8 |
| Madhya Pradesh | 43,873 | 35,018 | −20.2 | 8864 | 8068 | −9.0 |
| Maharashtra | 115,824 | 96,566 | −16.6 | 19,766 | 15,479 | −21.7 |
| Daman & Diu | 1987 | 1822 | −8.3 | 353 | 328 | −7.1 |
| Dadra & Nagar Haveli | 4047 | 4044 | −0.1 | 594 | 594 | 0 |
| Goa | 2856 | 2806 | −1.8 | 544 | 460 | −15.4 |
| Andhra Pradesh | 70,860 | 68,577 | −3.2 | 12,018 | 11,232 | −6.5 |
| Karnataka | 44,970 | 41,600 | −7.5 | 8137 | 7815 | −4.0 |
| Kerala | 16,275 | 16,052 | −1.4 | 3295 | 2946 | −10.6 |
| Tamil Nadu | 72,712 | 68,140 | −6.3 | 11,728 | 10,436 | −11 |
| Puducherry | 1929 | 1847 | −4.3 | 319 | 300 | −6 |
| Lakshadweep | 22 | 22 | 0 | 6 | 6 | 0 |
| Bihar | 11,621 | 10,007 | −13.9 | 2073 | 1659 | −20 |
| DVC | 15,045 | 13,777 | −8.4 | 2206 | 2046 | −7.3 |
| Jharkhand | 5651 | 5482 | −3.0 | 1012 | 1012 | 0 |
| Orissa | 20,429 | 20,368 | −0.3 | 3505 | 3468 | −1.1 |
| West Bengal | 33,349 | 32,715 | −1.9 | 6162 | 6112 | −0.8 |
| Sikkim | 361 | 361 | 0 | 100 | 99 | −1 |
| Andaman & Nicobar | 220 | 165 | −25 | 40 | 32 | −20 |
| Arunachal Pradesh | 463 | 394 | −14.9 | 101 | 85 | −15.8 |
| Assam | 4992 | 4669 | −6.5 | 971 | 937 | −3.5 |
| Manipur | 522 | 464 | −11.1 | 118 | 115 | −2.5 |
| Meghalaya | 1405 | 1225 | −12.8 | 294 | 284 | −3.4 |
| Mizoram | 332 | 282 | −15.1 | 76 | 70 | −7.9 |
| Nagaland | 543 | 485 | −10.7 | 118 | 110 | −6.8 |
| Tripura | 813 | 735 | −9.6 | 220 | 197 | −10.5 |
| India | 783,057 | 715,795 | −8.6 | 120,575 | 108,212 | −10.3 |

### 1.6.1.1 Coal

It is the most widely available fossil fuel in India. Approximately 55% of India's energy demand is fulfilled by coal. The industrial heritage of India has been built upon the domestic coal. In the last forty years, the commercial primary energy consumption has risen by 700%. Taking into account the limited reserves of petroleum & natural gas in the country, the conservation of ecology constraints on hydel projects as well as the political viewpoint of nuclear energy, coal is set to take up the center stage in India's energy strategy.

There are 27 significant coalfields of coal deposits, which are located in the eastern and the south-central parts of India. There are 36 billion tons of lignite reserves across the country, with the southern State of Tamil Nadu accounting for 90%.

India's installed electric generating capacity grew from 1.7 GW in 1950 to 356 GW in March 2019. Of that total, roughly 55% is coal-fired. With this tremendous growth, India turned into both the world's third largest coal producer and importer by the end of 2018.

Coal capacity in the most drought-stricken areas—estimated to total 37 GW—must be addressed as a priority. A prediction has been made that at present rate of utilization the remaining reserves of coal will last only about the next 80 years and it can last for over 140 years if all the inferred reserves are also considered at present rate of extraction. However, if India will try to meet the energy demand, the consumption of coal would also increase and thereby the reserves will be finish before the forecasted year.

If the production of coal rose at a rate of 5%, the extractable reserve would be emptied by the next 45 years. However, it is highly sturdy to estimate the requirement for coal because of the massive change in costs as well as the availability of other substitute fuels and their technological advancements. A significant issue about coal is that its deposits are highly focused in the Eastern regions of India. To set up a coal-fired power plant in the Western or North-western areas in India would mean transportation of coal through distances longer than 1000 km and because of this longer distance, the cost of coal-powered energy will become unfavorable.

By the financial year 2017, the coal deficiency in India rose to 400 million tons from 50 million tons as in 2011 as per a report by Credit Suisse. Further, a report of the Planning Commission of India shows that demand for coal is expected to increase from around 937 million tons by 2021–22 and to more than 1415 million tons by 2031–32. This would force the power either generating companies to go for other alternatives such as acquiring offshore coal, through mine acquisitions or buying coal from international markets.

"Technically, the dependence on imported coal is not viable as old power stations cannot take the heat generated from more than 10%–12% international coal blending." Mine acquisition is not easy for India, as most of the deals have been sealed by China, leaving India with fewer alternatives. Besides, the problem with mining abroad, similar to India, is facing the issues of government and the environment clearance. Also, countries which are rich in resources like Indonesia, Australia, and South Africa are considering giving priority to the local demands. Indonesia has already executed the domestic market obligation and has restricted the export of coal since 2014. The transportation of the imported coal to the power plants is a significant issue as the country is lacking in proper infrastructure like ports and rail network. Rail capacity is a cause of concern, but trebling the size of ports to 3 billion from 1 billion in 2015 would improve the situation.

The cost of generating power through domestic coal is INR 2.1 per kWh. But the cost rises to INR 3.6 per kWh if imported coal is used, which is because of the high prices of international coal, port handling charges, and customs duty. This creates an issue for the power generating industry to maintain the utilization of capacity at high levels. Due to their poor financial conditions, the state electricity boards are unable to buy power, and therefore, India continues to suffer from power outages. The environment ministry has given support to the power industry by decreasing the number of "no-go" areas (places where mining is not allowed). However, still, the dependence of the sector on imported coal is set to remain high in the future years. More investment is also required to develop the infrastructure.

### 1.6.1.2   Petroleum/Oil

The total reserves of crude oil in India are around 1201 million metric tons. Crude oil production in India decreased to 662 BBL/D/1K in August from 667 BBL/D/1K in July of 2019. Several factors contribute to the long-term growth in demand of petroleum products like economic growth (GDP), the impact of energy conservation measures, etc.

Road infrastructure plays a crucial role in increasing the requirements of petrol and diesel, and its price. The growth of alternative transportation modes and the development of alternatives fuels such as biofuels and technologies like hybrids are highly desirable. The production of automobiles has dramatically increased in the last decade in India, as indicated in Table 1.3. Due to rapid growth of automobiles sector, the demand for petroleum products will witness a growth in order, and it is expected to rise to more than 240 million metric tons by 2021–22, which will further increase to around 465 million metric tons by 2031–32 considering high output growth.

**TABLE 1.3**
**Production of Automobiles in India**

| Year | Car Production | Commercial | Total Vehicles Production | % Change |
|------|----------------|------------|---------------------------|----------|
| 2018 | 4,064,774 | 1,109,871 | 5,174,645 | 8 |
| 2017 | 3,952,550 | 830,346 | 4,782,896 | 5.83 |
| 2016 | 3,677,605 | 811,360 | 4,488,965 | 7.90 |
| 2015 | 3,378,063 | 747,681 | 4,125,744 | 7.30 |
| 2014 | 31,62,372 | 6,82,485 | 38,44,857 | −1.40 |
| 2013 | 3,155,694 | 742,731 | 3,898,425 | −6.60 |
| 2012 | 3,296,240 | 878,473 | 4,174,713 | 6.30 |
| 2011 | 3,040,144 | 887,267 | 3,927,411 | 10.40 |
| 2010 | 2,831,542 | 725,531 | 3,557,073 | 34.70 |

### 1.6.1.3 Natural Gas

The natural gas production is 32.65 BCM in the year 2017–18. The percentage of growth in natural gas production is 2.36. Existing fuels in many sectors for feedstock and energy can be very easily replaced by natural gas. Although this alternative is going to depend on the cost of gas when compared to various other fuels. Thus, the demand for gas would depend upon the natural gas price relative to that of alternatives (naphtha which is used in fertilizers and petrochemicals and coal for power).

India imports around 20.1 million tons of LNG per year and has started to find other substitutes like shale gases. While the domestic production is more than 140 million standard cubic meters per day, which is less than half of the demand, major issues in usage of natural gas are unfavorable rates and the fiscal policies adopted for oil and gas. The oil produced domestically has international price, whereas the government puts a cap on the price of natural gas at very lower levels. It is because of highly subsidized Indian fertilizer industries, which is the dominant user of natural gas.

### 1.6.1.4 Nuclear Energy

The fourth largest source of electricity is nuclear energy in India. As of March 2018, India has 22 nuclear reactors in operation in 7 nuclear power plants, with a total installed capacity of 6780 MW. India has a low reserve of Uranium. Only 10,000 MW can be fueled with the available uranium supply in the pressurized heavy water reactors (PHWR). By 2050 the aim is to supply 25% of electrical energy from nuclear power.

Although the trade ban stopped India importing uranium, it helped it to develop indigenous technology and trained a large workforce in the nuclear field. It is anticipated that foreign technology and fuel will boost India's nuclear power plants considerably with the signing of a nuclear cooperation agreement. However, the initial euphoria has somewhat ebbed as India failed to sign the Nuclear liability bill. "Nuclear Power Corporation India Limited" is responsible for generating and maintaining nuclear power plants, and it has planned to build five nuclear power plants having a capacity of eight reactors of 1000 MW. Future nuclear power plants are planned at locations such as Kudankulam in Tamil Nadu, Jaitpur in Maharashtra, Mithi Virdi in Gujarat, Haripur in West Bengal, and Kovvada in Andhra Pradesh.

In spite of its enormous benefits, power obtained from nuclear energy is very costly. Although in support of this kind of power generation it can be said that the initial capital cost is huge, the subsequent investments are not that large. Further, nuclear power faces a considerable amount of resistance from the local population as its associated dangers are large, and damage caused recently to nuclear power plants in Japan happened due to an earthquake, and a subsequent tsunami has increased the fears among them. India lies within the zone of seismic activity. India has a very secretive nuclear establishment, and therefore very few debates on its safety and security are done in public. This is compounded by the fact that the Atomic Energy Review Board functions under the administrative control of the government and is not fully independent.

The National Disaster Management Authority (NDMA) has done a "gap analysis," which was presented to the federal home ministry in 2009. Twelve significant vulnerabilities were discussed which would hamper the response of country if any disaster strikes on even one of its seven nuclear power plants. Issues pinpointed includes an inadequacy of medical personnel who have been trained in handling injuries related to radiation, lack of alternative sources of food and water, absence of shelters during emergency periods, absence of facilities of camping near nuclear plants, inadequate participation of home guards, police and civil defense volunteers as first responders and a lack of emergency response centers. This document highlighted "the stock of monitoring equipment and personnel protection gear is minimal and needs to be augmented to upgrade the capabilities to handle nuclear emergencies."

Recently, leakage from a disused gamma irradiator occurred at a scrap market in west Delhi, which killed one person and injured five others. Medical personnel were not able to identify the symptoms of radiation exposure at that time.

The only countermeasure so far has been the formation of 10 National Disaster Response Force (NDRF) squad numbers over 8000 personnel. They have been skilled by Bhabha Atomic Research Centre (BARC), and they are operating under the orders of the ministry of home affairs. These squads are located in places such as Delhi, Pune, Ernakulum, and Kolkata are pretty well armed with the necessary accessories to respond to any such emergency in the future, but that still is not sufficient to cope with a severe nuclear accident as occurred in Japan in 2011.

## 1.7   ENERGY SECURITY IN INDIA

The primary role of energy security in any country is to cut the dependency on imported energy sources for its economic growth. India is set to experience a shortfall in the energy supply. The gap has gradually increased since 1985, as India became a net importer of coal. It has not been able to increase the production of oil by a significant margin since the 1990s. Increasing oil demand about 10% per annum has led to a huge import bill. Besides, there is a significant monetary loss to the government as the government gives a subsidy on refined oil product prices.

The dependence on importing of energy is estimated to rise massively in the future. The oil imports have met with 75% of oil consumption demands. The importing of gases like LNG is estimated to rise in the future. The dependence on energy import indicates susceptibility to the external price shocks and supply fluctuations, and it threatens the energy security of the country.

All these plans could be plausible; but their implementation is going to take its time. For developing countries such as India, over reliance on reserves is set to be slow due to its constraints in the resources. Besides, the market has not been yet developed enough, and the monitoring agencies are not at all experienced enough to predict the supply situation to take necessary action. Lack of storage capacity is one of the primary causes of worry, and it has to be adequately seen if our country has to make an increment in its energy stockpile. Amongst all alternatives, the simplest and the most readily available option to reduce the demand by implementing permanent conservation of energy.

## 1.8 FUTURE ENERGY STRATEGY

The future energy strategy can be divided into immediate, medium-term, and long-term action plans/strategies. The types of action plan/strategies are as follows:

### 1.8.1 IMMEDIATE TERM ACTION PLAN

- Justify the price list of energy products.
- Optimum usage of existing resources.
- Reduction in distribution losses as well as efficiency of production systems, including traditional sources of energy.
- Promotion of R&D, use of technologies, technology transfer, and practices for environmentally friendly energy systems.

### 1.8.2 MEDIUM TERM ACTION PLAN

- Demand management through proper energy conservation, by optimizing the fuel mix, and creating a relevant model for the transport sector. A higher dependency on railway systems rather than road is required in the transportation of goods and passengers. Also required are shifting from private to public passenger transportation and changes in the design of different products to decrease the material density, reuse, recycling, etc.
- A shift to less energy-intensive types of transport has become an absolute necessity. This primarily includes to enhance the transport infrastructure such as roads and optimum design of vehicles. Compressed natural gas (CNG) and synthetic fuel are better scheme for public and private transport. Smart city scheme implementation will be able to decrease the demand for energy use in the transport sector.
- It is essential to shift from non-renewable to renewable energy resources such as solar, wind, biomass energy, etc.

### 1.8.3 LONG-TERM ACTION PLAN

- *Efficient production of energy resources*
  - Efficient generation of energy resources like coal, oil, and natural gas.
  - Decrease of flaring of natural gas.
- *Enhancing energy* infrastructure
  - Installing new refineries.
  - Developing urban gas T & D network.
  - Maximizing the rail transportation efficiency for production of coal.
  - Constructing new coal and gas-fired power stations.
- Improving energy efficiency
  - Enhancing efficiency in consonance to priorities such as national, socio-economic, and environmental.
  - Encouraging the efficiency of energy and its emission standards.
  - Marking programs for products and adopting energy efficient technologies in large-scale industries.

- Liberalizing and privatizing the energy sector.
    - Decreasing cross aids on oil products and energy tariffs.
    - Uncontrolled coal prices while making competitive prices of natural gas.
    - For improving efficiency, oil, coal, and power to be privatized.
- Legislation of investment in order to appeal to foreign investments.
    - For appealing the private sector participation, making the approval process easy in power generation, transmission, and distribution.

Economic-evaluation methods simplify comparisons in various energy technology investments. Guidelines are needed by all sectors of the energy community in order to make economically efficient energy-related decisions. The same technique can also be used for comparing investments in energy supply or energy efficiency.

## QUESTIONS

Q.1. Explain in brief the following:
    a. Renewable and non-renewable energy.
    b. Commercial and non-commercial energy
    c. Low-grade and high-grade energy
    d. Energy security
Q.2. Write a note on the Indian energy scenario.
Q.3. State the need for energy conservation in India concerning our present scenario.
Q.4. List at least five states in India where the coal deposits are highly concentrated.
Q.5. What is the percentage of India's oil consumption which is imported and what is the approximate cost per year?
Q.6. Mention places of any three oil reserves in India.
Q.7. What is the potential of hydropower generation which is available in India and how much has it been utilized thus far?
Q.8. Mention the percentage shares of consumption of commercial energy in the industrial as well as in the agricultural sectors.
Q.9. Co relates the economic growth of a country with energy consumption.
Q.10. What are your views on the long-term action plans which are essential for the long-term management of energy in India?
Q.11. Write a note on the subsidies in oil sectors in India.
Q.12. Discuss in brief. The different reforms in the energy sector.
Q.13. How $CO_2$ is regarded as a potential threat to the planet even when the respiration and decomposition of plant release more than ten times $CO_2$ when compared to the amounts which are being released by human activities?

# 2 Energy Management and Energy Conservation Opportunities

## 2.1 ENERGY MANAGEMENT

The objective of energy management is to generate goods and deliver services with the minimum cost and less impact on environment. The process of energy management is not only managing the generated energy from various sources of energy (fossil fuel, nuclear, solar, wind, biomass, etc.) but also optimum use of energy in energy-consuming devices. However, energy management is defined as "the judicious and effective use of energy to maximize profits at minimum cost and enhance competitive positions." The action plan of regulating and optimizing energy and using systems and procedures to minimize the required energy per unit of output while holding constant or minimizing the total cost of manufacturing output from systems is known as energy management. The aim of energy management is to accomplish and maintain optimum energy procurement and consumption, throughout the institution and:

- To reduce the cost of energy/waste without affecting quality and production capacity and
- To reduce environmental impacts.

### 2.1.1 ENERGY MANAGEMENT TECHNIQUES

1. Self-information and consciousness amongst the people
2. Reengineering and assessment
3. Updates of technology

#### 2.1.1.1 Self-Information and Attentiveness Amongst the Crowds

The process of energy management is successful and the idea is implementing with the help of good information of process and machine. The operational process of machine is very important to know the leader. Initially, there is resistance from the user to adopt new concepts or technologies. This is due to the user's fixed perception toward energy consumption in their domain. This requirement should be identified and explained in the beginning itself. It is necessary to understand the cost and benefit of the energy conservation to the owner.

Awareness on the part of the proprietor can offer maximum economic and minimum cost solutions to save energy. Awareness on the part of the people can have up to 5% of energy saving.

**Example:**

1. Promoted the concept of

**"Zero Production = Zero Power Consumption"**

At present, it has become quite important to save electricity during unused time or idling period. One should understand this idea and shut down the system as well as the supporting devices.
2. After the microprocessor-based timer was introduced, all supporting system/auxiliary equipment auto switch off during idling times, which results in huge energy saving.

### 2.1.1.2   Reengineering and Estimates

We need to explain in detail about minimum costing and awareness. The scope and range of energy conservation in the field of consideration are required to be determined.

We need to determine the present condition and employed technology in the form of process requirements, manufacturing capacity, and ability. Sometimes, we are handicapped with the ability and capacity of machine for the sake of energy saving. It need not be necessary to complete the work on time. The potential of energy optimization will occur when it is established. We require to start assessment and reengineering of the process or machine or electricity consumption equipment. It may be changed or replaced the layout, capacity of motor, starters, and type of load, etc., for the successful implementation of the reengineering process.

### 2.1.1.3   Updates of Technology

Once the energy-saving opportunities are recognized in the specific area, the appropriateness, sustainability, and pricing are required to be analyzed for the latest available technology. Economics like payback period, investment cost return, amount of energy savings, etc., should be determined.

**"BETTER THE DIAGNOSIS, BEST WILL BE THE RESULT."**

## 2.2   TARIFF

Tariff is defined as the rate of electrical energy supplied to a customer. The cost of electricity generation depends on the magnitude of electricity customers and their load. There are various kinds of consumers such as industrial, domestic and commercial; therefore consideration should be given at the time of fixing tariff.

This helps in fixing suitable rates to highly complicated problems.

### 2.2.1   Objectives of Tariff

The electrical energy is sold at various rates and various ways; therefore, it not only recovers the generation cost but also makes a judicious profit. The tariff should contain the following factors:

1. The costs of electricity generation from the power station.
2. The cost of capital investment for the transmission and distribution system.

3. The cost of operation and maintenance of supply of electricity, i.e., metering device, billing, etc.
4. Making suitable benefit on the capital investment.

## 2.2.2 CHARACTERISTICS OF A TARIFF

Electricity tariff should have the following necessary features:

1. **Proper Return**: The tariff should be such that it ensures the proper return from each consumer. Other way, the tariff is the sum of revenues from the customers that is equal to the cost of electricity generation and transmission, and distribution plus judicious profit. This will allow   confirmed continuous and consistent supply of electricity from firm to customers.
2. **Fairness**: The tariff must be reasonable to facilitate their various types of customers are fulfilled with the charges of electricity. Hence, a large user customer should be charged at a lesser cost as compared to a smaller one. That is due to increased electricity consumption extend the fixed rate over a larger number of units, therefore decreasing the overall rate of generating electrical energy. Also, a customer whose load variation do not turn greatly from the ideal (i.e., constant) must be rated at a lesser charge as compared to one whose load variation change significantly from the ideal.
3. **Simplicity**: It must be simple and easily understandable to an ordinary customer. The complex tariff may cause an obstruction from the community that is usually distrustful of transmission or supply companies.
4. **Reasonable Profit**: In the tariff, the profit segment should be judicious. The companies of electrical energy supply are a community service company and enjoy the profit of monopoly. Hence, the investment is comparatively safe due to noncompetition in the market. Due to this factor, the profit is limited up to 8% or so per year.
5. **Attractive**: If the tariff is attractive then the number of electricity-using customers is increased. Try to make a simple and easy way to fix the tariff so that customers can pay without difficulty.

## 2.2.3 VARIOUS KINDS OF TARIFF

There are various types of tariff. The commonly used tariffs are the following types:

1. **Simple Tariff**
   A fixed rate per unit of consumption of energy is known as a simple tariff or uniform rate tariff.
   In simple tariff, the charge of price per unit is constant, which means it does not change with rise or reduction in the number of units consumed. Energy meters record the amount of electrical energy that is consumed at customer's terminals. This is the simplest among all tariffs and is easily understood by the consumers.

**Demerits**

a. Simple tariff has no discernment between various types of customers because every customer has to pay equal charges, which means that the rate of charge is fixed.
b. The rate per unit distributed is high.
c. The use of electricity is not encouraged.

2. **Flat Rate Tariff**

When various types of customers are charged at dissimilar uniform per unit charges, this is known as a flat rate tariff.

In flat rate tariff, the customers are grouped into several classes and every class of customers is charged at a various uniform rate. For example, may be 60 paise is the flat rate per kWh for lighting load. However, it was possibly slightly less (let us consider 55 paise per kWh) for power load. The various classes of customers are made with consideration of their diversity and factors of load. The merits of this tariff are fairer to several types of customers and moderately simple in calculations.

**Demerits**

a. Individual meters are needed for lighting load, power load, etc., as flat rate tariff varies according to the method of supply used. Therefore, the application of flat tariff is costly and complicated.
b. An individual class of customers is charged at the same cost irrespective of the amount of electricity used. However, large-energy-consuming customers are charged at a lesser rate as in this situation the fixed rate per unit is decreased.

3. **Block Rate Tariff**

The charge of energy in a given block at a specified rate and the consequent blocks of energy are rated at gradually reduced rates, this is known as block rate tariff.

In this tariff, the energy utilization is divided into blocks and the rate per unit is secured in every block. The rate per unit in the first block is maximum and it is steadily reduced for the consequent blocks of electricity. For example, 3.00/kWh may be charged for the first 200 units; 2.50/kWh may be charged for the next 200 kWh; the remaining additional units may be fixed at the rate of 2.00/kWh.

The advantage of block rate tariff is the customer takes an incentive to use more electricity. This raises the load factor of the system and hence the price of power production falls.

Therefore, the major defect of this tariff is lacking a measure of the customer's demand. The block rate tariff is being used for larger number of domestic and small commercial customers.

4. **Two-Part Tariff**

If the rate of electricity is charged on the basis of the highest demand of the customer and the unit used, this is known as two-part tariff.

In this tariff, the total charge to be considered from the customer is divided into two constituents, viz., fixed rate and variable rate. The fixed rate depends upon the highest demand of the customer whereas the running charges depends upon the number of units used by the customer.

Therefore, the customer is charged at an assured amount per kWh of highest demand plus assured amount per kWh of electricity used and the expression is calculated as:

$$\text{Total charges} = \text{Rs.}\,(b \times kW + c \times kWh)$$

where:

b is the rate per kW of highest demand and

c is the rate per kWh of electricity used.

This category of tariff is generally valid to industrial consumers due to considerable high demand.

**Merits**

a.  It is a simple tariff and customers can understand easily.

b.  It recovers the fixed rate that depends upon the highest demand of the customer but is independent of the units used.

**Demerits**

a.  The customer has to pay the fixed rate regardless of the fact whether he has used or not used the electricity.

b.  There is always an error in calculating the highest demand of the customer.

5. **Maximum Demand Tariff**

It is like a two-part tariff with the only variance that the highest demand is actually measured by installing highest demand meter in the premises of the customer. It eliminates the hostility to two-part tariffs where the highest demand is evaluated only based on their respective rate. This category of tariff is generally useful to big user. Nevertheless, it is not suitable for a small electricity consumer (e.g., domestic customer) as a separate highest demand meter is necessary.

6. **Power Factor Tariff**

When power factor of the customer's load is considered in tariff calculation, it is called power factor tariff.

The role of power factor tariff is an important in an alternating current (AC) system. A minimum power factor maximizes the rating of station equipment and line losses. Thus, a customer having minimum power factor essentially penalized. Some of the important types of power factor tariff are following:

a.  **kVA maximum demand tariff**: It is the revised form of two-part tariff. In this circumstance , the fixed rate is created on the basis of highest demand in kVA and not in kW. As kVA is inversely proportional to power factor. Hence, a customer having low power factor has to pay more toward the fixed rate. The advantage of this tariff is that it inspires the customers to run their appliances and equipment at better power factor.

b.  **Sliding scale tariff**: The average power factor tariff is called sliding scale tariff. In this tariff, the average power factor, about 0.8 lagging, is used as a reference. If the power factor of the customer decrease below this factor, appropriate additional rates are made. On the other way, a concession is permitted to the customer, when the power factor is beyond the reference.

c. **kW and kVAr tariff:** Here, supplied active power (kW) and reactive power (kVAr) are rated individually. A customer having less power factor will draw added reactive power and therefore will have to pay higher rates.

7. **Three-Part Tariff**

When the sum of rate to be recovered from the customer is divided into three parts, viz., fixed rate, semi-fixed rate, and running rate, this is called as three-part tariff.

For example:

$$\text{Total charge} = \text{Rs.}\left(a + b \times kW + c \times kWh\right)$$

where a is the fixed charge made during every billing duration.

It contains interest and depreciation on the secondary delivery cost and labor cost of collecting revenues,

where:

b is the charge per kW of maximum demand and

c is the charge per kWh of energy used.

By addition of a fixed charge or customer's charge to two-part tariff, we obtain the three-part tariff. The main objection raised against this type of tariff is that the charges are divided into three constituents. This type of tariff is usually utilized by big users.

## 2.3   ENERGY CONSERVATION

Energy is an essential requirement of established economy and social configurations. One of the main problems regarding its supply is the exhausting nature of the withdrawal of fossil fuel, combined with the requirement for changeover to renewable energy resources. The last is contingent on a number of scientific and technological innovations. Energy conservation has the ability to fulfill the gap between supply and demand. Various steps for energy conservation are essential for consideration. Hence, conservation of energy is consuming less energy more judiciously than before. The supply of a watt is approximately always costlier than saving a watt. The power plant is one of the most capital-intensive industries. The resource of energy used efficiently is not only saving of energy but also saving of capital investment. Therefore, energy conservation is indeed the most economical energy "resource" until its potential is exhausted. The idea of energy conservation is dynamically changing. At present, the cost of fuel has increased and waste has become expensive but it was expensive to recover the waste energy in the past. The meaning of energy conservation is inhibition from wasting energy. For example: switch off lights instead of regular use, moderate cooling in the rooms with air-conditioners, and improvement in energy use efficiency through technological enhancement.

Energy conservation means enhancing the efficiency of energy consumption throughout society without compromising the comfort levels. Conservation of energy can be achieved through the following actions:

- Reducing the use of energy—It can be attained by reducing the consumption of energy, water, and other natural resources. For example, switch off lights in vacant rooms, brush teeth with a glass of water as an alternative to the wastage of water running from tap, use a pressure cooker for cooking food, etc.
- Use of energy-efficient technology—Application of energy-efficiency-certified home appliances such as light bulbs, cooking stoves, refrigerators for food storage, washing machines, furnace for heating, etc., which consume less power for their services.

## 2.4 ENERGY CONSERVATION OPPORTUNITIES (ECOs)

Opportunities to conserve energy are broadly classified into three categories:

1. **Minor ECOs**: It is simple, easy to implement, and needs less implementation time. It is human nature to be not alert to avoid careless wastage.
   Example: Stoppage of leakage points, lapses in housekeeping, and maintenance, etc.
2. **Medium ECOs**: Medium ECOs are more complex, and need extra investment and reasonable execution time. It may be due to change of technology: low power consumption compact fluorescent lamps (CFLs) replaces the bulb—although the purpose is the same (i.e., to provide light) the difference is lower energy consumption for the same illumination.
   Example: Replacement of current domestic appliances through new energy-efficient equipment.
3. **Major ECOs**: These offer substantial energy saving. They are a bit more complex, and require major investment and long implementation periods.
   Example: Replacement or significant renovation of old buildings, machinery, etc.
   Following are the sequence of steps to be followed for energy management and conservation in various sectors.

### 2.4.1 ENERGY CONSERVATION IN HOUSEHOLDS

1. **Lighting System**
   a. The light switch is one of the best energy-saving devices. Turn off lights when not essential.
   b. Various types of auto-run devices can help in saving energy used in lighting, by the use of infrared sensors, weight sensors, motional sensors, auto timers, dimmers, and solar cells relevant to switch on/off lighting circuit. Lighting load may connect to solar photovoltaic system to conserve conventional energy resources.

   c. Task-based lighting systems may be used, which concentrate light where it is desired. For example, a reading lamp illuminates only reading material.

   d. Regularly keep tube lights and lamps clean as dust can absorb 50% of the light.

   e. LED and CFLs convert electricity to visible light up to 5 times more than ordinary bulbs and also save approximately 70% of electricity for the same lumen level.

   f. An incandescent lamp converts 90% of the energy consumed into heat instead of visible light. Therefore, replace it by LED or CFL.

   g. It is estimated that 15 W CFLs give light equivalent to the 60 W ordinary filament bulbs.

2. **Room Air Conditioners**

   a. In summer, use of ceiling fan and table fan should be the first line of action in place of air conditioners because of economy and less power consumption.

   b. Proper design of shading over window and facade in homes may reduce upto 40% of air-conditioning load. Landscaping protects direct solar radiation on the wall, which also reduces cooling load.

   c. It is estimated that air-conditioner energy consumption is reduced by 3%–5% per degree temperature when set above 22°C. Therefore, thermostat of room air conditioner should be set at 25°C to provide the most thermal comfort with least expenses.

   d. A well-designed air conditioner cools and dehumidifies a room within 30 minutes; therefore, use a timer and switch off air conditioner for some time.

   e. Doors and windows should be air tight, which is easy to implement.

   f. Air-conditioner filters should be cleaned regularly. A dusty air filter obstructs airflow and may cause impairment in the AC unit. A clean filter helps the AC unit to cool down quickly with less energy consumption.

   g. Sometimes old air conditioners need to be repaired frequently and are expected to be inefficient; therefore, replace with energy-efficient air conditioners.

3. **Heating and Cooling Systems**

   a. Replace or clean or wash furnace and air conditioner filters routinely to retain energy efficient systems.

   b. Clean the filter of AC that can clear the air passage; system will work efficiently.

   c. Curtains, hangings, carpets, and furniture should not block air vents. Vents should be cleaned frequently either with vacuum cleaner or manually (broom).

   d. Use only programmable thermostat-based systems to regulate heating and cooling.

   e. Fit ceiling fans for proper air circulation, which enhances the efficiency of heating and cooling systems.

   f. Put film on windows which are not in use generally.

g. Deciduous trees should be planted outside house on the south side (Northern hemisphere). It gives a shade in summer and allows sunlight in winter season.

h. Water-heated radiators are installed in a house or apartment for room heating purposes. The insulation board should be placed between the radiators and the outside walls to reduce its losses.

i. Try to avoid use of water beds, which consume huge amounts of energy to heat in the winter. If it is essential in extreme cold climates, then insulate and cover it to retain the heat inside.

j. Apply attic insulation to improve the efficiency of furnace and air conditioner.

4. **Water Heaters**

a. Reduce heat loss; always insulate hot water pipes or hot steam pipes, particularly where they are passing over unheated areas. No need to insulate plastic pipes.

b. About 18% of the energy consumption can be reduced by changing 60°C to 50°C peak temperature of hot water heater.

5. **Microwave Ovens and Electric Kettles**

a. Microwaves conserve energy by dropping cooking times. They can save approximately 50% of energy consumed in conventional cooking, particularly for small amounts of food.

b. Microwaves cook the food at the outer edge from the center of the food. If you are cooking more items simultaneously, then large and thick items should be kept outside.

c. Thermostat automatic shut-off button-based electric kettles are more energy efficient than electric/gas cook top to heat water.

d. Clean the electric kettle regularly to remove dirt as it consumes more electricity. Boiling mixture of water and vinegar is the best way to remove deposits.

e. Heat only the amount of water that is required for one drink. Otherwise there will be energy loss.

6. **Computers**

a. Switch off home/office equipment (computer, printer, scanner, etc.) when not in use. A computer that runs 24 hours a day consumes more energy as compared to an energy-efficient medium capacity refrigerator.

b. If a computer must be left on to run simulation, then turn off the monitor. The monitor itself consumes more than half the system's energy.

c. Laptop, computers, monitors, scanner and copiers are to be set in sleep mode. It cuts energy costs by approximately 40%.

d. Remove the pull of battery chargers for laptops, cell phones and digital cameras when not in use. It saves lots of energy.

e. There is misconception about screen savers. Screen savers save computer screens, not energy. Start-ups and shutdowns do not use any extra energy, nor are they hard on your computer components.

f. Always shut down computers when work is finished; it reduces system wear and saves energy. It reduces system wear and saves energy.

7. **Drying**

Wet clothes should hang either on clothesline or on a clothes tree stand to dry for some time. It disinfects the cloth under natural sunlight and saves electricity. In the winter, this also acts as natural humidifier in a hot and dry room.

a. Ironing the clothes immediately after drying the cloth reduces ironing time as well as saves energy.

b. The lint trap of washing machine which is associated with the drainage pipe should be empty after every use of the dryer

c. Light and heavy clothes should be dried separately for efficient use of the dryer.

d. Mount a dryer vent hood where dryer discharges to minimize heat loss from this hole.

8. **Batteries**

a. Always use a rechargeable battery and its charger.

b. Avoid using battery-operated toys.

9. **Cooking**

a. Use a microwave instead of an oven, a range or a toaster oven whenever possible.

b. Prefer small appliances over big ones, such as a toaster oven, electric teapot, rice cooker, electric frypan or a crockpot.

c. Always cover the pan while cooking the food.

d. Turn off the gas burner or oven before the food is completely cooked to save energy.

e. Always use a pressure cooker for making food.

f. Cook more food than required for one meal and then heat the leftovers in a microwave.

g. Always prefer glass or ceramic utensils for cooking food in oven.

h. Stainless steel utensils should be used for making chapatti-like food.

i. Keep those utensils clean that reflect heat during cooking.

j. Always use appropriately sized burners according to the bottom area of utensil.

k. Excluding pastries, cakes or similar bakery items, preheating of oven is not required.

10. **Large Purchasing Decisions**

a. Verify the functionality of the appliances from check meters before its replacement.

b. Always check the energy efficiency rating ("Energy Star" label) on appliances before purchasing.

c. Select energy-efficient washing machines, which consume less energy.

d. Choose dryers which have moisture sensors that automatically turn off the machine when clothes are dry.

e. Avoid using automatic ice makers, which consume more energy. Side-by-side refrigerators also consume more energy than a typical model.

f. Induction cook tops are the most energy-efficient cooking stove.

    g. Try to use or buy green energy, which is non-polluted, like solar or wind or biomass energy.

    h. Replace old windows with energy-efficient types.

    i. Prefer ecofriendly natural gas furnaces.

    j. Choose those appliances which consume less energy and are more efficient.

    k. Use light colored roof shingle/tiles to avoid heat absorption.

11. **Winter Opportunities**

    a. Cover air conditioners during the winter to reduce drafts.

    b. Always wear slippers and light sweaters so that a few degrees temperature of hot air blower could be lowered.

    c. Cover legs and/or torso with a lap quilt or blanket when at home.

12. **Summer Opportunities**

    a. Air-conditioner thermostats should be set at 25°C (77°F) or higher for the most energy-efficient operation.

    b. Temporary layers of semitransparent plastic film may be put on window glasses to block excess solar radiation; this reduces cooling load of air conditioner, and also the cost. It can be removed during winter season for passive heating.

    c. Indoor heat from cooking appliances can be reduced by either keeping these items outside or replacing with microwave oven.

    d. Air blowing from a fan inside a room gives a feeling 5 degrees cooler than the actual ambient temperature.

    e. Use window shades on sunny side walls or retain drapes closed or add room-darkening shades to obstruct the light/heat from the sun.

    f. The outdoor unit of a central air conditioner should be kept clean and clear from dust, debris and grass clippings. This dirt can splatter onto the unit in heavy rain and may block the air passage; hence, it should be inspected regularly.

    g. Trees or shrubs to shade the outdoor unit of air conditioner reduce electricity consumption. But they should not block the airflow.

    h. Try to avoid frequently opening doors and windows during the noon time in hot summer days. It permits cool air to escape and hot air to enter the room.

    i. Change the timing of energy-intensive activities such as laundry and dishwashing to off-peak energy-demand hours. This will increase electricity reliability during heat waves.

    j. The radiant barriers should be installed on roofs to reduce summer heat gain as well as cooling costs. Basically, it is a reflective material that reflects radiant heat rather than absorbing.

    k. The attic should have proper ventilation to release additional summer heat.

## 2.4.2 ENERGY CONSERVATION IN THE TRANSPORT SECTOR

1. There are number of guidelines to save fuel in the transportation sector under energy conservation act. Convenient and short travel plans have potential to conserve the fuel, which also restricts undesirable driving

2. Try to use public transport wherever possible and appropriate in place of personal vehicle. This is the most economical way for people to travel more distances with less fuel.
3. If someone has a choice of car and motorcycle/scooter, try to use the motorcycle/scooter up to two persons need to travel. Use car if more than two persons have to travel or heavy luggage is required to be transported.
4. Car-pooling is the most effective way of fuel conservation in commuting the same place.
5. Traffic jams can only be avoided by following traffic rules, which also saves huge amounts of fuel.
6. The use of good-quality lubricant also helps in fuel conservation.
7. There is a lot of wastage of fuel in sudden speeding, braking and stopping, clutch riding, idling, over-speeding and over-loading; hence, right driving behavior is important for fuel conservation.
8. Timely servicing and tuning of the vehicle makes it more energy efficient. It helps in fuel conservation and emission control.

### 2.4.3   ENERGY CONSERVATION IN THE AGRICULTURE SECTOR

1. Renewable energy resources should be used in various agricultural activities. For example: solar PV or wind-driven pumps, solar drying and dual fuel engine powered by biogas or producer gas or biodiesel. By this way cultivation will switch toward lower-carbon energy sources.
2. Application of energy-efficient pump sets and water-frugal farming methods in irrigating crops conserves huge amount of energy.
3. Energy use for traction can be reduced by using energy-efficient equipment or by lessening the need for traction through low-tillage agriculture.
4. Chemical fertilizers are energy-intensive products. Hence, there is a need for reduction in the use of chemical fertilizers. It can be achieved either by effective utilization or promoting organic/microbial fertilizers. Reduction in demand of chemical fertilizer will decrease energy use in the chemical industry.
5. Practice conservation tillage systems. It decreases labor time, the number of passes and equipment wear and tear; it improves soil aggregation for root establishment, improves water availability, stores carbon in the soil through retention of crop residue and saves fuel consumption around 3.5 gallons per acre. In all, the total energy required to grow a crop is reduced.
6. Enhance the efficiency of post-harvest drying and storage through use of better equipment and appropriate maintenance.
7. Reduce post-harvest food grain losses. It will meet the food needs and reduce the need for crop production. Thus, saved energy would be used in fresh production.

### 2.4.4   ENERGY CONSERVATION FOR THE INDUSTRIES SECTOR

1. **Thermal Utilities**
   a. Waste heat can be used for preheating the combustion air.

   b. Use variable-speed drives on large boiler combustion air fans for variable flows.

   c. If allowed, then burn waste.

   d. Insulate hot oil tanks.

   e. Always clean nozzles, strainers and burners, etc.

   f. Use proper temperature of oil for oil heaters.

   g. To lessen heat loss, close burner air and/or stack dampers when the burner is not in work.

   h. Renovate oxygen trim controls (5% reduction in excess air increases boiler efficiency by 1%)

   i. Automate/optimize boiler blowdown and recover heat during boiler blowdown process. This may warm the back-up boiler.

   j. Check the sediment and scale in the water side. One-millimeter thickness of scale deposition on the water side can increase 5%–8% fuel consumption.

   k. Clean regularly soot, fly ash, and slag on the fire side. Three-millimeter-thick soot deposition on the heat transfer surface can increase fuel consumption up to 2.5%.

   l. Optimize boiler water treatment and deaerator venting.

   m. Installation of economizers to preheat boiler feed water using exhaust heat saves significant amount of fuel.

   n. Recycle steam condensates.

   o. Optimize the most efficient approach for operating multiple boilers by studying the part-load characteristics and cycling costs.

   p. Multiple or modular boiler units should be taken into account instead of one or two large boilers.

   q. Launch a boiler-efficiency maintenance program. Initiate with an energy audit and follow up; later, continue the boiler-efficiency maintenance program under the energy management program.

2. **Steam Systems**

   a. Repair steam leaks and condensate leaks in the steam system because 3 mm diameter hole on a pipeline transporting 7 kg/cm² steam will waste 33,000 liters of fuel oil in a year.

   b. Collect work orders for repairing steam leaks which cannot be repaired during the heating season in view of system shutdown requirements. Mark each such leak with a durable tag with good description.

   c. Install back pressure steam turbines to create lower steam pressures.

   d. Apply more efficient steam-desuperheating approaches.

   e. Confirm that process temperatures are properly controlled.

   f. Maintain minimum possible acceptable process steam pressures.

   g. Try to minimize hot water wastage to drain.

   h. Eliminate or blank off all unnecessary steam piping.

   i. Make sure condensate is returned or reused in the process. A 6°C rise in feed water temperature by economizer/condensate recovery is equivalent to a 1% fuel consumption saving in boiler.

   j. Preheat boiler feed water.

    k. Ensure boiler blowdown recovery.

    l. Inspect operation of steam traps.

    m. Eliminate air from indirect steam using equipment. A 0.25-mm-thick air film creates the equivalent resistance to heat transfer as a 330-mm-thick copper wall.

    n. Examine steam traps frequently and repair faulty traps immediately.

    o. Recovery of vent steam must be taken into account, for example, large flash tanks.

    p. For water heating, use waste steam.

    q. Utilize absorption chillers to condense exhaust steam before returning the condensate to the boiler.

    r. Use electric pumps in place of steam ejectors if cost benefits permit.

    s. Launch a steam-efficiency-maintenance program. Initiate with an energy audit and follow up; later, continue the steam-efficiency-maintenance program under the continuous energy management program.

3. **Furnaces**

    a. Use doors or air curtains to stop infiltration of air.

    b. Check excess air to the optimum level and monitor $O_2/CO_2/CO$ levels.

    c. Improve burner design, combustion control and instrumentation.

    d. Confirm that the furnace combustion chamber is under slightly positive pressure.

    e. Use ceramic fibers in the case of batch operations.

    f. Match the load to the furnace capacity.

    g. Retrofit with heat-recovery devices.

    h. Examine cycle times and decrease the same.

    i. Provide temperature controllers.

    j. Make sure that flames do not touch the stock.

4. **Insulation**

    a. Restore damaged insulation. A bare steam pipe of 150 mm diameter and 100 m length transporting saturated steam at 8 kg/cm$^2$ would waste 25,000 liters of furnace oil in a year.

    b. Insulate all hot or cold metal.

    c. Regularly change the insulation whenever it gets wet.

    d. Use temperature measurement infrared guns for cold wall areas during cold weather or hot wall areas during hot weather.

    e. Make sure that all insulated surfaces are clad with aluminum.

    f. Insulate all couplings, valves and flanges.

    g. Insulate open tanks; 70% heat losses can be avoided by floating a layer of 45-mm-diameter polypropylene balls on the surface of 90°C hot liquid/condensate.

5. **Waste Heat Recovery**

    a. Retrieve heat from engine exhausts, drying oven exhausts, engine cooling water, hot flue gas, low pressure waste steam, boiler blowdown, etc.

    b. Retrieve heat from incinerator off-gas.

    c. Use waste heat for outside air heating, fuel oil heating, boiler feed water heating, etc.

    d. Chiller waste heat use to preheat hot water.

    e. Practice heat pumps.

    f. Practice absorption refrigeration.

    g. Use air-to-air exchangers, run-around systems, heat pipe systems and thermal wheels.

## 6. Electrical Utilities

### a. Electricity Distribution System

- Optimize the tariff structure with utility provider.
- Schedule your operations to maintain a high load factor.
- Shift loads to off-peak times, if possible.
- Minimize maximum demand by tripping loads through a demand controller.
- Stagger start-up times for equipment with large starting currents to minimize load peaking.
- Use standby electric generation equipment for on-peak high load periods.
- Correct power factor to at least 0.90 under rated load conditions.
- Reposition transformers close to main loads.
- Set transformer taps to optimum settings.
- Disconnect primary power to transformers which do not serve any active loads.
- On-site electric generation or cogeneration must be taken into account.
- Surplus electricity of captive power plant can be fed to grid.
- Cross-check utility electric meter with a personal meter.
- Shut off computers, printers and copiers at night when not required.

### b. Motors

- Properly size to the load for optimum efficiency. Energy-efficient motors offer 4%–5% higher efficiency than conventional motors.
- Use energy-efficient motors where economical.
- Use synchronous motors to improve the power factor.
- Check alignments.
- Provide proper ventilation. Motor life is estimated to be halved for every 10°C increase in motor operating temperature over the recommended peak.
- Check for under-voltage and over-voltage conditions.
- Balance the three-phase power supply. An unbalanced voltage can reduce 3%–5% motor input power.
- Demand efficiency restoration after motor rewinding. The incorrect motor rewinding decreases efficiency by 5%–8%.

### c. Drives

- Use variable-speed drives for large variable loads.
- Use high-efficiency gear sets.
- Use precision alignment.
- Check belt tension periodically.
- Eliminate variable-pitch pulleys.

- Use flat belts as alternatives to v-belts.
- Use synthetic lubricants for large gearboxes.
- Eliminate eddy current couplings.
- Shut off drives when not needed.

d. **Fans**
- Use smooth, well-rounded air inlet cones for fan air intakes.
- Avoid poor air flow distribution at the fan inlet.
- Minimize fan inlet and outlet obstructions.
- Clean screens, filters, and fan blades periodically.
- Use aerofoil-shaped fan blades.
- Minimize fan speed.
- Use low-slip or flat belts.
- Check belt tension regularly.
- Eliminate variable pitch pulleys.
- Use variable-speed drives for large variable fan loads.
- Use energy-efficient motors for continuous or near-continuous operation.
- Eliminate leaks in ductwork.
- Minimize bends in ductwork.
- Turn fans off when not needed.

e. **Blowers**
- Use smooth, well-rounded air inlet ducts or cones for air intakes to reduce losses.
- Minimize blower inlet and outlet obstructions.
- Clean screens and filters periodically.
- Optimize blower speed.
- Use low-slip or no-slip belts.
- Check belt tension regularly.
- Eliminate variable pitch pulleys.
- Use variable-speed drives for large variable blower loads.
- Use energy-efficient motors for continuous or near-continuous operation.
- Eliminate ductwork leaks.
- Turn off blowers, when not needed.

f. **Pumps**
- Operate pumping near best efficiency point.
- Renovate pumping to minimize throttling.
- Adapt to wide load variation with variable-speed drives or sequenced control of smaller units.
- Stop running both pumps. Add an auto-start for an on-line spare or add a booster pump in the problem area.
- Use booster pumps for small loads requiring higher pressures.
- Enhance fluid temperature differentials to reduce pumping rates.
- Repair seals and packing to minimize water waste.
- Balance the system to minimize flows and reduce pumping power requirements.

- Use siphon effect to advantage; in other words, don't waste pumping head with a free-fall return.

g. **Compressors**
- Consider variable-speed drives for variable load on positive displacement compressors.
- Use a synthetic lubricant, following the compressor manufacturer guidelines.
- Make sure that the lubricating-oil temperature is not too high (to avoid oil degradation and lowered viscosity starts) and not too low (to avoid condensation contamination).
- Change the oil filters regularly for proper functioning.
- Periodically inspect compressor intercoolers for proper functioning.
- Use waste heat from a very large compressor to power an absorption chiller or preheat process or utility feeds.
- Launch a compressor efficiency-maintenance program. Begin with an energy audit and follow up; then make a compressor efficiency-maintenance program as part of a continuous energy management program.

h. **Compressed Air**
- Coordinate multiple air compressors by installing a control system.
- Study part-load characteristics and cycling costs to judge the most energy-efficient mode for operating multiple air compressors.
- Cast aside over sizing; match the connected load.
- Load up modulation-controlled air compressors. It uses almost as much power at partial load as at full load.
- Switch off the back-up air compressor when not needed.
- Decrease the air compressor discharge pressure to the lowest acceptable setting. Reduction of 1 $kg/cm^2$ air pressure (8 $kg/cm^2$ to 7 $kg/cm^2$) will result in 9% input power savings. This will also reduce about 10% compressed air leakage rates.
- Set the highest reasonable dryer dew point temperature.
- Switch off refrigerated and heated air dryers when the air compressors are off.
- Use a control system to minimize heatless desiccant dryer purging.
- A control system should be used to reduce heatless dryer purging.
- Minimize accumulation of condensation, leakages, purges, and excessive pressure drops for energy efficient operation. A power loss of up to 0.5 kW results due to compressed air leak through a 1 mm hole size at a pressure of 7 $kg/cm^2$.
- To avoid regular air leaks from the drain, use proper drain controls.
- In order to reduce electric demand charges, use either engine-driven air compression or steam-driven air compression.
- Replace worn-out old V-belts with high-efficiency flat belts.
- When the major production load has been switched off, use a small air compressor.

- Air from the coolest location (but not air conditioned) should be used as air compressor intake. A 1% reduction in compressor power consumption is recorded for each 5°C reduction in intake air temperature.
- For heating building makeup air in winters, use air cooled after cooler.
- Make sure heat exchangers are not fouled with oil.
- Make sure to prevent the fouling of air/oil separators.
- Keep regular monitoring on pressure drops across suction and discharge filters. Clean or change filters immediately upon alarm.
- Compressed air storage receivers should be of proper size. Fully demulsible lubricants and an effective oil-water separator can minimize disposal costs.
- Substitutes of compressed air should be considered like blowers for cooling, hydraulic instead of air cylinders, electric instead of air actuators, and electronic instead of pneumatic controls.
- Instead of blowing with open compressed air lines, use nozzles or venturi kind instruments.
- Few rubber-type valves may leak nonstop after their age and crack, so regularly check leakage in drain valves on compressed air filter/regulator sets.
- Controlling of the packaging lines should be done with high-intensity photocell units rather than standard units with continuous air purging of lenses and reflectors in dusty environments.
- Launch a compressed-air-efficiency-maintenance program. Begin with an energy audit and follow up; then make a compressed-air-efficiency-maintenance program as part of a continuous energy management program.

i. **Chillers**
- If possible, raise the chilled water temperature.
- Use the least temperature condenser water available which the chiller can handle. A 5.5°C reduction in condensing temperature means a 20%–25% decrease in compressor power consumption.
- Raise the evaporator temperature. A 5.5°C increase in evaporator temperature reduces compressor power consumption by 20%–25%.
- Regularly clean the fouling in heat exchangers. A 1-mm scale formation on condenser tubes can increase energy consumption by 40%.
- Optimize the condenser and refrigerated water flow rates.
- Exchange old, inefficient chillers or compressors with higher-energy-efficient units.
- Use water-cooled instead of air-cooled chiller condensers.
- Energy-efficient motors are most suitable for continuous or near-continuous operation.
- Indicate suitable fouling factors for condensers.
- For coordinating multiple chillers, install a control system.

- In order to operate multiple chillers with maximum efficiency at the same time, a study of the part-load characteristics and cycling costs is to be performed.
- Running the chillers with the lowest energy consumption saves energy cost as well as fuels a base load.
- Avoid over sizing.
- Isolate the cooling tower as well as off-line chillers.
- Launch a chiller-efficiency-maintenance program. Commence with an energy audit and follow up; then make a chiller-efficiency-maintenance program as part of your continuous energy management program.

j. **Cooling Towers**
- Regulate the fans of the cooling towers with reference to leaving water temperature.
- Regulate the optimum water temperature, which has been determined from cooling towers and chiller performance data.
- If there are few fans, use two-speed or variable-speed drives for cooling tower fan control. In case of many fans, then stage the cooling tower fans with on-off control.
- When the load is less, shut down needless cooling tower fans.
- Prevent solar radiation by covering hot water basins to prevent algae growth. It reduces fouling.
- Clean the water distribution nozzles in cooling tower regularly.
- In order to obtain uniform water pattern, install new nozzles.
- Self-extinguishing PVC cellular-film fill should be used instead of splash bars.
- Square-spray ABS nozzles (practically non-clogging nozzles) should be used instead of old spray-type nozzles in the counter flow cooling towers.
- PVC cellular drift eliminator units are not only highly efficient but also have low pressure-drop and self-extinguishing characteristics. They are better than slat-type drift eliminators.
- The surrounding area of cooling towers must be free from all obstacles to air intake or exhaust.
- Optimize the angle of the cooling tower fan on a load and/or seasonal basis.
- Make corrections in the inferior fan balance as well as in the uneven and/or excessive fan blade tip clearance.
- Use a velocity pressure recovery fan ring.
- In hot weather, alter the direction of exhaust obtained from air-conditioned buildings to cooling towers.
- Repair or replace leaking lines in water basins of cooling towers.
- For appropriate functioning, analyze water overflow pipes.
- Consider the treatment of side stream water and optimize the chemical use.
- Prevent flow from huge loads to design values.

- Loads which are not operational should be shut down.
- Utilize blowdown water received from return water header.
- Blowdown flow rate should be optimized.
- Minimize he blowdown through automation.
- Direct blowdown for the other applications. Blowdown can be removed from anywhere; it is not necessary to remove only from the cooling tower.
- In order to minimize the buildup of ice, carry out winterization plan for cooling tower.
- During a lack of water flow, to prevent fan operation, interlocks should be installed.
- Develop a program for cooling tower efficiency maintenance. Begin with an energy audit; then continue with a program for cooling tower efficiency maintenance under an energy management program.

k.  **DG Sets**
- The load should be optimized.
- Use waste heat in order to produce steam hot water or provide power to an absorption chiller or other utility.
- For process needs, use jacket and head cooling water.
- Regularly clean air filters.
- For reducing DG set room temperature, exhaust pipes should be insulated.
- When capacity is greater than 1 MW, use heavy fuel oil, which is cheaper.

l.  **Buildings**
- Fill the outer cracks, gaps, or openings by using acrylic latex caulk, weather stripping, or gasketing, etc.
- Consider insulation of roof, windows, etc.
- Windbreaks should be installed near the external doors.
- Insulating glass should be used in place of single-pane glass.
- In the interior of the building, cover a few windows and the skylight areas with insulating wall panels.
- External windows may be replaced by insulated glass block when light is required but not visibility.
- Take advantage of landscaping.
- Use vestibules or revolving doors at primary exterior doors.
- Install self-closure on the doors. Use automatic doors, air curtains, strip doors, etc., at high-movement areas between conditioned and non-conditioned places.
- To reduce the building stack effect, install doors in stairways and vertical passages.
- In order to minimize the cost of HVAC and lighting, cleaning personnel should be brought during working hours or as soon after as possible.
- Apply dock seals at shipping and receiving doors.
- Prefer tinted glass, reflective glass, coatings, awnings, overhangs, draperies, blinds, and shades for sunlit exterior windows.

m.  **Water and Wastewater**
  – For lower-quality requirements, recycled water is suitable.
  – If the cost of sewer is dependent on the consumption of water, then recycle the water.
  – For reduction in the requirements of pump power, closed systems should be balanced.
  – Use the most economical types of water to fulfill the necessity.
  – Stop water leaks underground.
  – Inspect water overflow pipes for suitable functioning levels.
  – Minimize blowdown by use of automation.
  – Supply correct tools for washdown, particularly for self-closing nozzles.
  – Mount energy-efficient irrigation systems.
  – Decrease flows at water sampling stations.
  – Remove constant overflow from water tanks.
  – Immediately stop leakage in toilets and taps.
  – Use water regulators on showers and fix self-closing types of faucets in washrooms.
  – Use hot water of a comfortable temperature.
  – Install a smaller, energy-efficient hot water system to provide service hot water in the winter season instead of expensive and inefficient central heating systems.
  – Use solar water-heating systems for warm-water requirement.
  – If water must be heated electrically, consider accumulation in a large, insulated storage tank to minimize heating at on-peak electric rates.
  – Install multiple, distributed, small water heaters to curtail thermal losses in large piping systems.
  – Instead of using manual line bleeding, fix freeze protection valves.
  – Consider use of leased and mobile water treatment systems, particularly for deionized water.
  – For decreasing TOC and BOD surcharges, install pretreatment plant.
  – Confirm the readings obtained from water meters.
  – Confirm the flow from sewers if bills depend on it.

n.  **Miscellaneous**
  – If there is any unmetered utility then meter them. Find the root cause of deviation.
  – Any spare or idling equipment should be shut down.
  – Confirm that all the utilities that are unserviceable are switched off, especially cooling water or compressed air.
  – In order to properly coordinate various components (chillers, multiple air compressors, boilers, cooling tower cells, etc.), install automatic control.
  – Negotiate with the facilitator regarding utility contracts as per current loads and variations. Consider buying utilities from neighbors to handle peak demand.
  – Consider upgrading the low-bid inefficient equipment if the lease is set to extend for several years.

- In order to reduce the heat transfer in long pipelines, fluid temperature should be adjusted to within acceptable limits.
- Give regulatory orifices in purges.
- Cut down any unnecessary flow measurement orifices.
- Consider substitutes to high-pressure drops across valves.
- In summers, shut down winter heat tracing.

### 2.4.5  ENERGY CONSERVATION IN LIGHTING

- **Replacement of Conventional Fluorescent Lamps With Their Energy-Efficient Counterparts**
  Energy-efficient fluorescent lamps are manufactured based on tri-phosphor fluorescent powder technology. Not only do they have very good color-rendering properties but they also provide high luminous efficacy.
- **Incandescent Lamps versus CFLs**
  CFLs are highly rated as a replacement for the incandescent lamps, which have low wattage. Efficiency ranges from 55 to 65 lm/W while their average rated life is about 10,000 h and is 10 times longer than that of incandescent lamps. CFLs are recommended for building entrances, hotel lounges, bars, restaurants, living rooms, pathways, corridors, etc.
- **Mercury or Sodium Lamps versus Metal Halide Lamps**
  Mercury or sodium lamps provide a low color-rendering index when a comparison is made with metal halide lamps. Metal halide lamps provide efficient white light. Therefore, whenever high illumination levels are needed for color-critical or color-rendering applications then the metal halides are preferred. They are highly suitable for painting shops, inspection areas, assembly line, etc.
- **High-Pressure Sodium Vapor (HPSV) Lamps**
  HPSV lamps are used when color rendering is not critical. HPSV lamps are highly efficient. However, their color-rendering property is low. Therefore, HPSV lamp installation is suggested for yard lighting, street lighting, etc.
- **Filament Lamps versus LED Panel Indicators**
  LED panel indicator lamps are used extensively in manufacturing units for monitoring, fault indication, etc. Previously, for the same operations filament lamps were used.
  Disadvantages of filament lamps are as follows:
  - Energy consumption is high, about 15 W/lamp.
  - Filament lamp failure is high, having an operating life of about 1000 h.
  - Filament lamps are very sensitive to voltage fluctuations, and thus they are getting replaced by light emitting diodes (LEDs).
  
  Advantages of LEDs over filament lamps are as follows:
  - Power consumption is low (lesser than 1 W/lamp).
  - They resist voltage fluctuation.
  - They have a long operating life of about more than 100,000 h.
- **Distribution of Light**
  Energy efficiency cannot be obtained by mere selection of more efficient lamps alone. Efficient luminaries along with the lamp of high efficacy

achieve optimum efficiency. Mirror-optic luminaries with a high output ratio and bat-wing light distribution can save energy. For achieving better efficiency, luminaries that have light distribution characteristics appropriate for the task interior should be selected. The luminaries fitted with a lamp should ensure that discomfort glare and veiling reflections are minimized. Installation of suitable luminaries depends upon the height—low, medium, and high bay for heights less than 5 m, for heights between 5 and 7 m, and for heights more than 7 m, respectively.

System layout and fixing of the luminaries play a major role in achieving energy efficiency. This also varies from application to application. Hence, fixing the luminaries at optimum height and usage of mirror optic luminaries lead to energy efficiency.

- **Control of Light**

The simplest and the most widely used form of controlling a lighting installation is "On-Off" switch. The initial investment for this set-up is extremely low, but the resulting operational costs may be high. This does not provide the flexibility to control the lighting where it is not required.

Hence, a flexible lighting system has to be provided, which will offer switch-off or reduction in lighting level when not needed. The following light control systems can be adopted at design stage:

- **Group lighting system in order to provide greater flexibility in lighting control**

  Light system grouping can be controlled manually or by timer control.

- **Install microprocessor-based controllers**

  The control of lighting can be achieved by using logic units placed on the ceiling. It can take the pre-program commands and activate specified lighting circuits. Microprocessor-/infrared-controlled dimming or switching circuits control the lighting system significantly. Movement detectors or lighting sensors are fixed in advanced lighting control systems to send signals to the controllers.

- **Optimum usage of day lighting**

  Day lighting can be used in combination with electric lighting whenever it is allowed. This should not introduce glare or a severe imbalance of brightness in the visual environment. In many cases, a switching method, to enable reduction of electric light in the window zones during certain hours, has to be designed.

- **Install exclusive transformer for lighting**

  Lighting equipment faces a majority of problems due to fluctuations in voltage. Thus, it should be kept isolated from power feeders as it would provide better regulation of voltage. Efficiency of the lighting system is increased as this reduces the voltage related problems.

- **For lighting feeder, install servo stabilizer**

  This stabilizes the voltage for the lighting equipment. It provides an economical option to regulate voltage level fed to the lighting feeder. The functionality of chokes and ballasts will also improve because of stabilized voltage. In various industrial plants, voltage levels are high during non-peaking hours. Without any significant drops in the illumination levels, voltage levels can be optimized during this period.

- **High-Frequency Electronic Ballasts versus Conventional Ballasts**
  Less heat dissipation reduces the air-conditioning load, and up to 35% energy saving has been recorded. High-frequency (28–32 kHz) electronic ballasts have the following advantages over conventional magnetic ballasts:
  - Instant lighting
  - Better power factor
  - Can function in low voltage
  - Weight is less
  - Life of lamp is increased
  - Pros of high-frequency electronic ballasts outweigh the initial higher costs when compared to conventional ballast.

## 2.4.6 ENERGY-SAVING OPPORTUNITIES IN HVAC

- **Cold Insulation**
  Select a proper economical insulation thickness in order to minimize heat gains by all cold lines or vessels.
- **Building Envelope**
  Distinguish major areas for air conditioning using air curtains and false ceilings in order to optimize the air-conditioning volumes.
- **Minimizing Building Heat Loads**
  Use steps like roof cooling, roof painting with light colors, energy-efficient lighting systems, application of thermostat in air-conditioning systems, sun film applications, variable volume air system, etc., for minimizing air-conditioning loads.
- **Minimizing Process Heat Loads**
  Process heat loads are to be minimized by TR capacity as well as refrigeration level, i.e., temperature required by the following:
  - Increase in heat transfer area in order to accept high temperature coolant.
  - Avoid wastages such as heat gains, loss of chilled water, idle flows, etc.
  - Regular cleaning and descaling of heat exchangers.
  - Air flow should be optimized.
- **Refrigeration A/C Plant Area**
  - As per manufacturer guidelines, maintenance of all A/C plant components should be done regularly.
  - Close valves of idle equipment in order to avoid bypass flows.
  - Part load operations should be minimized by matching loads and plant capacity on line.
  - Implement variable-speed drives for variable process loads.
  - Optimize condenser and evaporator parameters regularly for lowest specific energy consumption and highest capacity.
  - Implement VAR systems where economics permit as a non-CFC solution.

## QUESTIONS

Q.1.   Define energy management. State the basic principles and benefits of energy management. List down the objectives of energy management.

Q.2.   Explain the stepwise procedure for assessing energy conservation opportunities in lighting system.

Q.3.   Explain the energy conservation by adopting following techniques:
a.   Replacing lamp sources
b.   Using energy-efficient light control equipment.

Q.4.   State the differences between energy conservation and energy efficiency.

Q.5.   What are the advantages of enforcing an energy efficiency program for industries?

Q.6.   Explain how energy conservation is vital in current energy scenario.

Q.7.   The Energy Conservation Act requires that all designated energy consumers get energy audits conducted by
(a) Energy manager (b) accredited energy auditor (c) managing director (d) chartered accountant.

Q.8.   Under the Energy Conservation Act, indicate the five designated consumers.

Q.9.   Mention any three main provisions of Energy Conservation Act 2001 which are applicable to designated consumers.

Q.10. List energy conservation opportunities in HVAC system.

Q.11. List energy conservation opportunities in households.

Q.12. List energy conservation opportunities in the transport sector.

Q.13. List energy conservation opportunities in the agriculture sector.

Q.14. List energy conservation opportunities in the industries sector.

# 3 Energy Audit

## 3.1 CONCEPT OF ENERGY AUDIT

The audit of energy is an idea behind a proper planned method for carrying out decisions in the field of energy management. Its main motive is utilizing the net energy supplied and using it to its maximum potential and finding out all the streams of energy in a facility. It considers energy as a quantitative entity and calculates it by separating it into various types of individual functions. An industrial energy audit is a useful tool for establishing and carrying out a complete program of energy management.

In the Conservation Act of Energy, 2001, the definition of energy audit was explained as "the verification, monitoring, and analysis of the use of energy including submission of technical report containing recommendations for improving energy efficiency with cost-benefit analysis and an action plan to reduce energy consumption."

Inside an enterprise, the most expensive aspects are energy (thermal and electrical), labor, and raw elements. If the management of cost is considered, then for each of the above aspects, energy would emerge as the highest in terms of savings and cost reductions possibilities. Hence, energy management is a critical area for reduction in the cost strategically. Energy audit is useful in energy and fuel utilization in the industry and also helps in pointing out the areas where there is a possibility of energy wastage occurrence and where there is a chance of improving.

Energy audit has a positive impact on the reduction of energy costs and the establishment of maintenance parts and to control the quality of the products, which are directly related to the utility of energy and production. These audit programs will help to keep track of any anomaly in the cost associated to the energy, energy availability, and reliability. By finding out the perfect match of a combination of different energy, pointing out the technologies related to energy conservation, and providing alternative better parts to equipment related to energy conservation.

In practice, the energy audit is a willingness to change ideas related to the conservation of energy into a reality, using technologically better solutions. To take considerations of the economic factors associated with organizational behavior over a particular period is important.

The primary aim of an energy audit is to find out various pathways for reducing the consumption of energy per unit of the output of the product or reducing the cost of operation. The energy audit provides a "Benchmark" (base point) in management of energy in an organization along with a primary platform for more efficient use of the energy in an organization.

## 3.2   TYPE OF ENERGY AUDIT

The energy audit is the initial approach in improving building efficiency and industrial facilities. The energy audit is generally divided into four types and they are as follows:

1. **Walk-through audit**: It is a small visit to the area of facilities for finding simple and inexpensive loopholes in energy use. And also, for improving its maintenance and operating cost and increase the savings associated with the operating cost.
2. **Utility cost analysis audit**: A utility cost analysis is the careful and detailed analysis of energy usage in the matter of quantity cost of operation of the facilities. Typically, several years of data of utility are studied to understand the variation in the usage of energy, maximum demand, the effect of weather, and energy-saving possibilities.
3. **Standard energy audit**: The standard energy audit is composed of a detailed analysis of the energy systems of the facility. To be specific, the standard energy audit contains a base value for the energy use of the facilities. Typically including, the energy savings, the appropriate cost associated with different measures of energy conservation.
4. **Detailed energy audit**: A detailed audit of energy is one of the most illustrated and a very time-consuming process. The main parts of this energy audit include instruments to measure the energy use for an entire building or any energy system inside the building (such as light, office equipment, fan, HVAC, coolers, etc.). On top of that, computer simulations are carried out for detailed energy audits to recommend new parts for the facility.

### 3.2.1   The Type of Energy Audit to be Performed Depends On

- Working and classification of enterprise/organization,
- Auditing details up to which it is needed, and
- Possibilities and magnitude of cost quantum desired.

Hence, energy audit is classified in two ways.

- Preliminary audit
- Detailed audit

### 3.2.2   Preliminary Energy Audit Methodology

A preliminary energy audit is a short and fast process:

- Establish the consumption of energy in the industry/enterprise.
- To find out the opportunity for cost/energy reduction.
- To find out relevant and the simplest areas for consideration.
- To find out the instant (especially no-/low-cost) modification/savings.

- Fix a point of reference.
- To find out areas for more descriptive and illustrated studies/calculations.
- A preliminary energy audit uses current data or the data that can be obtained easily.

### 3.2.3  Detailed Energy Audit Methodology

### 3.2.3.1  General Process for Detailed Audit of Energy

For carrying out an energy audit, few duties have to be done depending on the requirement needed for audit (type of audit). It depends on the size and working of such Audit. A few of the tasks may have to be repeated, reduced in scope, or even eliminated based on the findings of other tasks. Hence, carrying out an energy audit is not a straightforward method; it is nonlinear. However, a basic outline can be drawn by the following points:

**Step 1: Facility and Utility Data Analysis**: This step is carried out for finding out the energy systems characteristics and the pattern of energy usage in case of a building or any facility. The characteristics of such a building/facility can be understood using different types of mechanical/architectural/electrical drawings or by consulting with the building supervisor. The average energy usage can be found out from the last few years of bills of electricity and other utilities, which will also help in finding out any variations due to climatic/seasonal condition on the facility/building energy usages. The tasks which can be performed are as follows:
  - Finding utility data for a minimum of three years (for finding out the history of the energy-use pattern).
  - To find out the uses of different types of fuel and energy sources, etc. (identify fuel type which is responsible for highest energy usages)
  - Finding the variation of fuel usages with the type of fuel (for identifying maximum/peak demand for energy usage with the type of fuel).
  - To find out the utilization rate composition (energy and demand rates) (for finding if there is any penalization for high/peak consumption/demand of a building and purchasing of fuel can be made cheaper).
  - To find if any changes in weather create a variation in consumption of fuel (to identify any variation in energy usages corresponding to intense weather conditions).
  - Analyzing the utilization of energy for a building by taking into account of its type and size; the signature of the building can be found, which included per energy use unit area.

**Step 2: Walk-Through Survey (Overview via Survey)**: This method finds different approaches to save energy in terms of quantity.
  **Few tasks associated with these steps are**:
  - To find out the demand of the consumers and needs.
  - Existing working and proceedings of maintenance are to be checked.
  - To find current working conditions of large energy-using devices such as HVAC, electric motors, etc.

- Calculate the occupancy, equipment, and lighting to find out energy use density and operational hours.

**Step 3: Baseline for Energy Use in Buildings**: The working idea behind this process is to develop a base case model that shows the present building energy usage and operational status. This basic model acts as a reference for keeping an estimate of the amount of energy savings occurred after from few of the measure taken for energy conservation.

**The primary tasks to be performed during this step are as follows**:

- Procuring and reviewing the various engineering drawings (architectural, mechanical, electrical, and control drawings).
- Inspection, testing, and keeping an account of the current building equipment for determining the performance of the building based on their efficiency, and reliability.
- Keeping a track of all the working hours of the system/equipment and also account of any system which remains idle (lighting and HVAC systems).
- Create a baseline model for usage of energy in building.
- Check the baseline model using the data of utility and/or recorded data.

**Step 4: Assessment of Energy-Saving Measures**: This step is to reduce the effective cost of energy by doing different analysis. Energy savings and economic analysis are done to find the effective cost energy conservation measures. The procedures to save energy are as follows:

- Make a detailed list of energy conservation remedies using data collected during walk-through survey.
- Find out energy savings using numerous energy conservation measures relevant to the building by means of the baseline energy-use simulation model developed in "Step 3."
- Evaluate the initial costs essential to incorporate the energy conservation measures.
- Determine the cost-effectiveness of all recommended energy conservation measures by payback period or life cycle cost analysis.

A comprehensive energy audit provides a detailed energy project execution plan for any project because it takes into account the significant energy users. It provides an accurate idea of energy usage since it considers the collective impacts of all projects, energy use of all major equipment, and comprises of comprehensive energy saving cost and project cost.

Energy balance is the one of the crucial elements in a comprehensive audit. An inventory of energy consuming systems, assumptions of present operating conditions and energy use calculations are the basis of energy balance. Then, evaluated energy use is compared to the actual charges (utility bills).

There are three phases in detailed energy audit: Phase I, Phase II, and Phase III.

Phase I—Pre-Audit Phase
Phase II—Audit Phase
Phase III—Post-Audit Phase

### 3.2.3.2 A Guide to Energy Audits Conducting at a Glance

It will depend on the type of industry since the energy audit needs to be dynamic.

An extensive and illustrated 10-step procedure is given below for conducting an energy audit at the field level. Energy manager or energy auditor may change these steps based on his/her understanding and the type of industry (Table 3.1).

#### 3.2.3.2.1 Phase I—Pre-Audit Phase Activities

Pre-audit is a planned process for carrying and working out an efficient audit of energy. Hence, initially, a survey is to be carried out.

---

### TABLE 3.1
### Ten-Step Method for Conducting Detailed Audit of Energy

| Step No. | Action Plans | Reason/Result |
|---|---|---|
| 1. | **Phase I Pre-audit**<br>• Decide and chalk out the idea<br>• A pen and paper estimation<br>• Interview with energy manager of the plant | • Figuring out the resources and create a team for the energy audit.<br>• Figure out instrument and time-bound.<br>• Collection of data of the major events.<br>• Getting acquitted with plant activities.<br>• A self-assessment of the operation and practices are taking place. |
| 2. | • Meeting with the board of members and the divisional heads | • Developing cooperation<br>• Sets questions related to the audit for each department.<br>• Creating awareness and orientation. |
| 3. | **Phase II Audit phase**<br>• The first collection of data and flow chart of the process and utility of energy. | • Collecting and analyzing historical data and baseline data.<br>• A flow chart diagram of the process.<br>• Diagram of the service utilities.<br>• Creating a design and finding out operating data and the dates of the operations.<br>• Annual energy bill and the amount of consumption of energy. |
| 4. | • Carrying out surveys and also monitoring | • Measurements:<br>The motor survey, insulation, and lighting survey with portable instruments for better and accurate data collection. Check the design data with the operating data. |
| 5. | • Carrying out detailed paths for a selected energy guzzlers | • Trials/Experiments:<br>  • 24 h of power monitoring<br>  • Load variations trends in pumps, fan compressors, etc.<br>  • Trial for boiler efficiency<br>  • The trail for furnace efficiency<br>  • Performance of the equipment |
| 6. | • Analysis of the usage of energy | • Energy wastage analysis and balancing of energy materials analysis. |

*(Continued)*

**TABLE 3.1 (*Continued*)**
**Ten-Step Method for Conducting Detailed Audit of Energy**

| Step No. | Action Plans | Reason/Result |
|---|---|---|
| 7. | • Identification and development of energy conservation (ENCON) opportunities | • Identification and consolidation of ENCON measures.<br>• Develop and think and refine the ideas.<br>• Reviewing the ideas earlier given the unit personnel.<br>• Review the earlier ideas given by the energy audit if any.<br>• Use brainstorming and value analysis techniques.<br>• Getting better equipment of higher efficiencies from the vendor. |
| 8. | • Cost-benefit analysis | • Assess technical feasibility, economic viability, and prioritization of ENCON options for implementation.<br>• Select the most promising projects.<br>• Prioritize by low, medium, and long-term measures. |
| 9. | • Presentation and report to the higher most management | • Documentation, report submission, and presentation to the higher most management. |
| 10. | **Phase III Post-audit phase**<br>• Implementation and follow-up | Assisting and implementing recommendation measures of ENCON and performance monitoring.<br>• The action plan and the timeline for implementation.<br>• Follow-up and review periodically. |

### Initial Visit at the Site and Preparing for Detailed Audit

A visit to the site is to familiarize with the person-in-charge and also for the understanding of the site. It may take only few hours or maximum one day to judge the essential actions to complete the energy audit.

Following actions are to be carried out during the site visit:

- Consult with the site supervisor and plan the objectives of the energy audit.
- Examine the economic guidelines based on the recommendations of the audit.
- Examine the main energy consumption data along with the concern in-charge.
- To find out technical drawings and designs of the sites of building layout, various mechanical/electrical systems, etc.
- Walk-through of the area along with an engineer/in-charge.

### The reasons for the site visit are as follows:

- Create a team for the audit.
- Finding out critical areas of energy consumption to be inspected during the audit.
- Identifying existing and additional instruments required for the audit.
- Identify the necessity for instrument's to be installed before or during the audit. For example, kWh, oil or gas, steam meters.

- Scheme a time-bound frame.
- Data collection about the plant energy sources and the major consumption areas.
- Creating awareness through campaigns.

### 3.2.3.2.2   Phase II—Activities of a Descriptive or Detailed Energy Audit

This phase is a detailed study about the process, equipment, and other materials needed. A detailed or descriptive energy audit can take few weeks to some months in completion.

The reports include all input-output data of the various process of each section in the industry. The efficiency is being calculated for each of the processes and it is being monitored for improvement. There should be detailed recommendations regarding points where improvement is possible and then followed by a pre-formed implementation of those processes for justification.

Information to be collected for a detailed audit is as follows:

1. Different types of energy consumption based on department, equipment, process of product, etc.
2. Material balance data (all materials that directly or indirectly related to the process). For example: supplied raw materials, intermediate product stage and final products, use of waste products or scrap, recycled materials, manufacturing of by-products for reuse in other industries, etc.
3. Energy cost and tariff data.
4. Process and material flow diagrams.
5. Generation and distribution of site services. For example: compressed air, process steam, etc.
6. Source of energy supply. For example: electricity from the grid or self-generation through captive power plant, etc.
7. Feasibility of fuel replacement, process improvement, and application of co-generation systems (combined heat and power generation).
8. Organizing the energy management and conservation awareness program in the industry.

Current baseline data and annual reports are used for understanding consumption, production, and productivity level in view of the input raw materials. The following baseline data are to be collected:

- Existing technology, operations used, and equipment features.
- Capacity utilization.
- Quantity and type of input materials used.
- Consumption of all types of resources. For example: water, fuel, electricity, etc.
- Other input parameters. For example: compressed air, cooling water, etc.
- Quantum and variety of wastes generated.
- Quantity of products being rejected.
- Efficiencies/yield.

## 3.3   COLLECTING DATA STRATEGY

An essential part of the data collection is to handle additional data strategically. Some of the essential tips for data handling are illustrated as follows:

- Easily understandable systems of measurement and also accuracy in terms of requirement are needed.
- Using inexpensive measuring devices (measuring the flow rate using conventional methods).
- Acquired data should be detailed and qualitative for accurate gist (the grading of the product, the average production, etc.).
- To find out the period, after which the data collection is to be done again for process variations.
- Measurement workouts over anomalous workload periods. For example: startup and shutdowns time workload.
- For difficult measurement areas, design values are to be used. For example: measurement at cooling water through a heat exchanger.

Prepare a flowchart which represents all the essential process operations and usage of material and energy by using designs, drawings, and other records.

At the same time, numerous inputs and output streams at each process step should be elucidated by energy audit team. For example: A flowchart of Penicillin-G manufacturing is given in the Figure 3.1. Note that waste and other consumption have been identified.

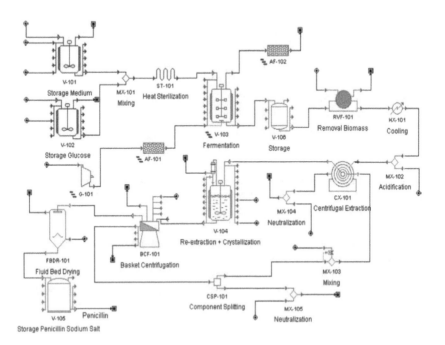

**FIGURE 3.1**   Flowchart of Penicillin-G manufacturing. (From Brunet R. et al., *J. Cleaner Prod.*, 76, 55–63, 2014.)

The focus areas of the audit are the waste generation, resources consumption, raw materials input, and efficiency of the process. The process illustrates the possibilities of significant energy conservation areas.

### Recognizance of Energy Conservation Opportunities

**Fuel substitute**: Finding the best fuel for the specific energy efficient conversion.

**Generation of energy**: Finding out the efficiency enhancement opportunity for the energy conversion system/equipment/utility, for example, captive power generation, steam boilers, thermic fluid heating, optimum DG set loading, gasifier, cogeneration systems, etc.

**Energy distribution**: Identifying the scope to improve the efficiency from the technology of the network, such as cables, switch gears, improvement in power factor, chilled water, cooling water, compressed air, etc.

**Energy usage processes**: Here is the maximum opportunity of improvement, most of these are cryptic. Dividing the energy in terms of process and analyzing each process and optimizing it.

## 3.4  TECHNICAL AND ECONOMIC FEASIBILITY

The following issues illustrate the technical feasibility.

- Accessibility of technology, skilled labor, reliability, and workspace.
- The effect of energy-efficient measures on production, safety, and quality of products.
- The maintenance and availability of the spare parts for the facilities.

Economic feasibility often becomes a critical parameter for management acceptance. The economic analysis is carried out in a variety of procedures. Example: Payback method, internal rate of return method, net present value method, etc. The payback method is the simplest and adequate method for small investment short duration measures. Worksheet model for estimating economic viability is given in Table 3.2.

**TABLE 3.2**
**Sample Worksheet for Economic Feasibility**

Name of Energy Efficiency Measure

| 1. Investment | 2. Annual Operating Costs | 3. Annual Savings |
| --- | --- | --- |
| Equipment | Cost of Capital | Thermal Efficiency |
| Civil Works | Maintenance | Electrical Energy |
| Instrumentation | Workforce | Raw Material |
| Auxiliaries | Energy, Depreciation | Waste Disposal |
| **Net Saving/Year (Rs./Year) =** | | **Payback Period in Month =** |
| **(Annual Saving – Annual Operating Cost)** | | **(Investment/Net Saving/Year) × 12** |

## 3.5   TYPES OF ENERGY CONSERVATION MEASURES

There are some feasible energy saving projects based on energy audits and analyses of the industry. They are divided into three groups:

1. Low cost—High return
2. Medium cost—Medium return
3. High cost—High return

Generally, the low cost–high return projects are of greater importance. Rest of the projects has to be investigated, engineered, and priced for execution in a phase wise manner. Generally, projects associated with energy cascading and process changes comprise high costs along with high returns. Therefore, such projects need careful scrutiny before funds commitment. These projects are generally difficult and may take a long time in execution. Project priority guidelines are given in Table 3.3.

## 3.6   UNDERSTANDING ENERGY COSTS

The sense of energy cost is an essential criterion for energy cost reduction. Industries not having enough energy metering devices should use fuel and electricity bills. The annual company balance sheet is another useful tool in such cases.

Energy invoices can be used for the following purposes:

- Bills/ invoices provide complete data of annual energy used and purchased, and gives a baseline for future reference.
- Energy cost data may show the potential for savings when linked to various activities/operations in the plant.
- Purchasing of electricity based on maximum demand tariff.
- Suggestions can be made when savings is possible.
- Energy bills can be utilized to compute the energy and cost savings after implementation of energy conservation measures.

## TABLE 3.3
### Project Priority Guideline

| Priority | Economical Feasibility | Technical Feasibility | Risk Feasibility |
|---|---|---|---|
| A Good | Well-defined and attractive | Existing technology adequate | No-risk/Highly feasible |
| B Maybe | Well-defined and only marginally acceptable | Existing technology may be updated, lack of confirmation | Minor operating risk/ Maybe feasible |
| C Held | Poorly defined and marginally unacceptable | Existing technology is inadequate | Doubtful |
| D No | Not attractive | Need breakthrough | Not feasible |

### 3.6.1 Fuel Costs

Many types of fuels are available for thermal energy supply. Some are listed as follows:

- Fuel oil
- Low sulphur heavy stock (LSHS)
- Light diesel oil (LDO)
- Compressed natural gas (CNG)
- Biogas
- Producer gas
- Liquefied petroleum gas (LPG)
- Coal
- Lignite
- Biomass

It is very easy and simple to understand the fuel cost and procurement in tons or kiloliters. One should always consider the three major features such as availability, cost, and quality at the time of purchasing. The factors are illustrated as follows:

- Price at origin, cost of transportation, and type of transportation.
- Fuel quality (contaminations, moisture, etc.).
- Energy content (calorific value).

### 3.6.2 Power Costs

In most of the country, the price of electricity differs a lot, starting from city to city and consumer to consumer. The factors involved in deciding the final purchase cost are as follows:

- Energy charges at the maximum demand, kVA. How fast the electricity is consumed?
- Energy charges, kWh. How much electricity is consumed?
- Time of day (TOD)/Time of use (TOU) charges, peak/non-peak period. When electricity is utilized?
- Power factor charge (P.F.) and real power use vs. apparent power use factor.
- Another incentives and penalties imposed from time to time.
- Changes in high and low tension tariff rate.
- Cost of slab rate and its variation.
- Type of tariff clause and rates according to different categories such as residential, industrial, agricultural, commercial, etc.
- Tariff rate in developed and underdeveloped areas.
- Benefits in tax for new projects.

## 3.7 BENCHMARKING AND ENERGY PERFORMANCE

Internal (historical/trend analysis) and external (across similar industries) benchmarking of energy consumption are the two most useful tools for performance assessment and logical evolution of pathway for reformation. Daily/monthly/yearly trend analysis of energy consumption, cost, appropriate manufacture features, and specific energy consumption, support to know the effects of capacity utilization on energy use efficiency and costs on a wider scale.

External benchmarking is related to the unit comparison among the unit of a similar group. Moreover, the critical part is to find a match among the groups; or else findings would be misleading. There are few comparing elements to be considered during benchmarking externally:

- Operation scale
- The development of technology
- Quality and specifications of raw material
- Quality and specifications of production

Benchmarking performance of energy allows:

- Quantifying the fixed and variable energy consumption trends (Production level).
- Comparing the performance of energy across each production level in the industry (capacity utilization).
- Finding out the best method using benchmarking data.
- Finding possibilities and areas of energy consumption and cost reduction.
- The platform for observing and setting target exercises.

The benchmark parameters can be:

- Related to gross production.
  **For example**: Production of cement in kWh/MT; production of yarn in kWh/kg; production of paper in kWh/MT or kCal/kg; power production in kCal/kWh; production of urea or ammonia million kilocals/MT.
- Related to equipment and utility.
  **For example**: Refrigeration by air conditioning plant in kW/ton; thermal efficiency of a boiler in %; cooling tower effectiveness in % cooling; generation of compressed air in kWh/NM$^3$; power production based on diesel in kWh/liters.

The benchmarks are needed for mentioning the meaningful comparisons among other manufacturers. Such cases are given as:

- Cement manufacturing plant—types of cement, Blaine number, i.e., process used, and Portland is to be mentioned along with kWh/MT.
- Textile manufacturing—yarn types, count on an average, i.e., polyester/cotton used is to be informed along with kWh/square meter.
- Specific power consumption of compressed air—A comparison at similar inlet air temperature and generating pressure.
- Paper plant: Paper type, raw material (recycling extent), GSM quality is used to mention along with kWh/MT, kCal/kg.

- Power plant/cogeneration plant: Plant percentage loading, condenser vacuum, inlet cooling water temperature is used to mention along with heat rate (kCal/kWh).
- Fertilizer plant: Capacity utilization (%) and on-stream are the values compared while mentioning specific energy consumption.
- Air conditioning plant: Chilled water temperature and refrigeration load (TR) are the key factors for comparing along with kW/TR.
- Cooling tower effectiveness: Ambient air wet/dry bulb temperature, relative humidity, air, and circulating water flows are needed to be mentioned for comparison.
- Diesel power plant: Performance should be compared at similar loading (%), steady run condition, etc.

## 3.8  PLANT ENERGY PERFORMANCE

Plant energy performance (PEP) is the measurement of more or less energy used in plant for manufacturing the products based on the past data. It analyzes the good performance of energy management being carried out. A yearly energy consumption comparison is made to the current condition based on the production output. Plant energy performance monitoring set a reference year, and on that basis, the other years are compared to evaluate the progress. Since the production output varies yearly; hence, the energy required to produce the same thing based on the reference years is taken into consideration and is then compared. The value obtained determines whether there is any improvement or deterioration since the reference year.

### 3.8.1  PRODUCTION FACTOR

It is used to calculate the energy required to produce this year's production output if the plant is operated in the same way as in the reference year. It is the ratio of production of the present year to that of the reference production year.

$$\text{Production factor} = \frac{\text{Current Year's Production}}{\text{Reference Year's Production}}$$

### 3.8.2  REFERENCE YEAR EQUIVALENT ENERGY USE

The reference year's energy use is the energy use during the production of the present year, which is generally known as "reference year energy use" or "reference year equivalent" in brief.

The reference year equivalent can be calculated as:

$$\text{Reference year equivalent} = \text{Reference year energy use} \times \text{Production factor}$$

The enhancement or decline from the reference year is called "Energy performance," and it measures the development in plant energy management. It is the increase or decrease in the current year's energy use to the year of reference and it is measured by deducting the current year's energy use from the reference year's equivalent. Then it is divided by reference year equivalent and multiplied by 100 to get in percentage as:

Plant energy performance

$$= \frac{\text{Reference year equivalent} - \text{Current year's energy}}{\text{Reference year equivalent}} \times 100$$

Plant energy performance is the percentage of energy saved at the current rate of use as compared to the reference year rate of use. More the improvement means higher the number.

### 3.8.3 MONTHLY ENERGY PERFORMANCE

Sometimes in industry, instead of a yearly basis, monthly plant energy performance is carried out for controlling and monitoring energy usage during the entire process.

## 3.9 FUEL AND ENERGY SUBSTITUTION

Replacement of costly, polluting, and finite sources of fuel/energy with less expensive or freely available and less polluting resources. For example, substituting fossil fuel with less cost/less polluting fuel/energy (natural gas, solar photovoltaic, biogas and locally available agro-residues). Fossil fuel reliance can be reduced either by energy conservation or substitution, thereby it improves the economy also. Sometimes fuel substitution is done for the betterment of life. For example: Kerosene and Liquefied Petroleum Gas (LPG) have substituted soft coke in domestic application to reduce health problem. These kind of initiatives have been taken by many countries. Few examples of fuel substitution are:

- Use of coconut shells or rice husks instead of coal.
- Using LSHS instead of LDO.
- Natural gas is becoming the choice for fuel and feedstock in the fertilizer, petrochemicals, sponge iron industries, and power.

Few examples of energy substitution are:

- Using a steam heater instead of electric ones.
- Using solar water heating systems instead of steam-based hot water.

## 3.10 THE ENERGY CONSERVATION ACT, 2001 AND ITS FEATURES

### 3.10.1 POLICY FRAMEWORK—ENERGY CONSERVATION ACT—2001

The Energy Conservation Act, 2001 (Act No. 52 of 2001) was endorsed by the Govt. of India and later on amended as "The Energy Conservation (Amendment) Act 2010 (Act No. 28 of 2010) in August-2010." It provides the much-needed legal framework and institutional arrangement for embarking on an energy efficiency drive. In the domains of the Act, the Bureau of Energy Efficiency (BEE) is created on March 1, 2002, by merging the erstwhile Energy Management Centre of the Ministry of Power. The Bureau is responsible for execution of policy program and organization of implementation of energy conservation activities.

### 3.10.2 Important Characteristics of the Energy Conservation Act

The Standards & Labeling (S&L) program has been identified as a key thrust area of BEE. The S & L program would ensure that only energy efficient equipment and appliance would be made available to the customers. Following are the main provisions of the Energy Conservation Act on Standards & Labeling program:

- Put a barrier to the minimum energy consumption and performance standards for devices.
- Manufacturers are not allowed sale and import of any products which are not following the standards.
- Announce a compulsory labeling scheme for goods/equipment so that customers can make an informed choice.
- Publicize information on the profits to customers.

### 3.10.3 Designated Consumers

The foremost regulations of the Energy Conservation Act made for the designated consumers are:

- The government would inform about the industries and establishment which takes in a tremendous amount of energy as designated consumers;
- A list of energy-intensive consumer (railways, port trust, transport sector, power stations, transmission and distribution companies and commercial buildings or establishments) is supplied by Schedule to the Act;
- An energy audit to the designated consumers done by the accredited energy auditor;
- The designated consumers are needed to appoint qualified energy manager as per BEE guidelines; and
- Designated consumers should follow the energy consumption rules and regulations as prescribed in Energy Conservation Act/BEE.

### 3.10.4 Certification of Energy Managers and Accreditation of Energy Auditing Firms

The essential duties under this consideration as illustrated in the Act are:

A team of professionally eligible energy managers and energy auditors with proficiency in financing, implementation of energy efficiency projects, policy analysis, and project management would be established through certification and accreditation programme. A national-level certification examination is to be conducted by the BEE for the selection of energy managers and energy auditors, and also design and organize training modules.

### 3.10.5 ENERGY CONSERVATION BUILDING CODES

The key provisions of the Energy Conservation Act on Energy Conservation Building Codes (ECBC) are:

- ECBC guidelines are to be prepared by the BEE.
- A notification would be provided for the local climatic conditions and also other factors for commercial building by the individual states which are erected after the notification and the rules relating to energy conservation building code. Moreover, the buildings must have 500 kW of connected load or 600 kVA contract demand or above and to be explicitly used for commercial purposes.
- Energy audit will also be fixed for specific designated commercial building consumers.

### 3.10.6 CENTRAL ENERGY CONSERVATION FUND

Energy Conservation Act provisions in this case are:

The fund development at the center (Central Govt.) to create the delivery mechanism at nationwide adoption of energy efficiency facilities namely performance contracting and promotion of energy service companies. The purpose of the fund is for more focus in R&D and demo to fill the market with energy efficient equipment and appliances. It will help to generate the amenities for testing and development and to encourage consumer awareness.

### 3.10.7 BUREAU OF ENERGY EFFICIENCY (BEE)

The moto of BEE is to establish an energy efficiency services that enables facilities for delivering across the country. It acts as a leader in energy efficiency in all sectors of the economy. The main intention is to minimize the energy intensity in the Indian economy.

The general administration, guidelines and management of the affairs of the Bureau is vested in governing council with 26 members. The council is led by Power Minister and council members, who are secretaries of numerous line ministries, the CEOs of technical agencies under the ministries, equipment and appliance manufacturers, industry, architects, consumers and five power regions representing the states. The Bureau Director-General will be an ex-officio member-secretary of the council.

The Central government has to support the BEE initially to grant through a budget. The period of 5–7 years will be enough to become self-dependent. The authorization would be provided for collecting an appropriate fee on completion of the duties assigned to it. The BEE can also use the Central Energy Conservation Fund and other funds received from numerous sources for innovative financing of energy efficiency projects for encouraging energy efficient investment.

### 3.10.7.1    Role of Bureau of Energy Efficiency

The function of BEE is preparing standards and labeling of equipment and appliances, making a list of designated consumers, certifying and accrediting processes, preparing codes for buildings, maintaining the central energy conservation fund, and engaging with the states and the central agencies in activities of promotion. Also, the role of developing energy service companies (ESCOs), making the energy market with efficient products, and creating awareness by taking steps such as house cleaning.

### 3.10.7.2    Role of Central and State Governments

The following role of Central and State government is visualized with the Act:

- Central—to inform about the rules and norms under different provisions of the Act, support initially in financing funds to BEE and EC, contacting the state level governments to notify, enforce, penalize, and adjugated.
- State—to update/amend the energy conservation building codes according to the local/regional climate situations, to create an agency on the state platform for coordination, regulating and enforcing various provisions of the Act, and form a State Energy Conservation Fund to promote energy efficiency activities.

### 3.10.7.3    Enforcement Through Self-Regulation

Energy Conservation Act would only monitor two actions. The process is explained below for self-regulation and is to be carried out for the areas to be verified.

- Accredited energy auditor certifies the energy consumption norms and standards of the production process for implementing the energy efficiency in designated consumers.
- Manufacturer's declared values regarding energy performance and standards will be checked in accredited laboratories by picking sample from market. Any industry, consumer, or consumer associates have the freedom to question the declared quality of the product made by other industries and put to the notice of the BEE. The BEE is the association recognized for testing the challenged product in disputed cases to take measures for self-regulation.

### 3.10.7.4    Penalties and Adjudication

- Under such an Act, any violation/offence would mean a fine of Rs. 10,000 of every breach of a law and an extra of Rs. 1000 per day for continued non-compliance.
- There would be no penalties initially for the first five years as it would be for promoting and developing the infrastructure.
- The State Electricity Regulatory Commission has the power to adjudicate, and shall nominate an adjudicating officer to carry out an inquiry for imposed penalty.

## 3.11   RESPONSIBILITIES AND DUTIES TO BE ASSIGNED UNDER THE ENERGY CONSERVATION ACT, 2001

### 3.11.1   ENERGY MANAGER: RESPONSIBILITIES

- Formulate an annual plan of the activities to be taken place and introduce to management regarding financially appealing investments to decrease the energy costs.
- Launch an energy conservation cell inside company with the permission of management about the mandate and task of the cell.
- Develop activities to rectify monitoring and process control for reducing the energy consumption and cost.
- Examine equipment/machine/device performance with regard to energy efficiency.
- Making sure the proper working and calibration of the various measuring equipment so that correct energy usage can be determined.
- Develop materials containing energy conservation information and also conducting a workshop for members associated with the enterprise.
- Reform disintegrating of energy consumption data down to shop level or profit center of company.
- Develop method for precise measuring of specific energy consumption for the various goods or services or activity of company.
- Creating and handling training programs for energy efficiency at the level of operation.
- Help in the selection of management personnel for various outside programs.
- Develop an information center on sectoral, nationwide, and international development on energy efficiency technology and management system and information value.
- Develop a united system of energy efficiency and environmental improvement.
- Initiate execution of energy audit and energy efficiency improvement projects with external agencies collaboration.
- Interact with the other energy managers of different firms and organizations for information exchange.

### 3.11.2   ENERGY MANAGER: DUTIES

- Submit annual report in BEE format regarding energy consumed and action taken as per direction of the accredited energy auditor to BEE and state level designated agency.
- Develop better data handling systems such as recording, collection, and analysis to monitor energy consumption.
- Backing to accredited energy audit team hired by the company for conducting energy audit.
- Send information to the BEE regarding the task given by a mandate and the job description as per the Act.

- Develop a plan of efficient energy utilization, energy conservation, and implementation with consideration of economic stability of the investment as per rules and regulations of the Energy Conservation Act.

### 3.11.3 ENERGY AUDITORS: RESPONSIBILITIES

- Once in a year, conduct an internal audit of the unit/section/system/equipment individually.

### 3.11.4 ENERGY AUDITORS: DUTIES

- Submit reports to the energy manager along with a recommendation for the actions.
- Maintain records of calibration repots of all energy measurement equipment.
- Keep up to date portable tools/instruments required for audit.
- Place equal all codes of practices for energy efficiency test.
- Be a part as team member of external audit team.
- Be the project verifier in ESCO performance projects for the M&V system and baseline and savings.

## 3.12 ENERGY AUDIT INSTRUMENTS

For carrying out an energy audit, amount of energy and the type of energy needs to be verified; hence, it requires measurements. Instruments are needed for carrying out such measurements. Instruments must be very much flexible in nature such as portable, handy, and easy to use and inexpensive. The parameters normally measured during energy audit are:

The elementary electrical parameters in alternating current (AC) and direct current (DC) systems are voltage (V), current (I), power factor, active power (kW), apparent power (demand) (kVA), reactive power (kVAr), energy consumption (kWh), frequency (Hz), harmonics, etc.

There are some more important parameters other than electrical such as temperature, heat flow, radiation, air and gas flow, liquid flow, revolutions per minute (RPM), air velocity, noise and vibration, dust concentration, total dissolved solids (TDS), pH value, moisture content, relative humidity, flue gas analysis $CO_2$, $O_2$, CO, SOx, NOx, combustion efficiency, etc. The operating manuals of the instruments should be understood sincerely. The energy audit team member should be familiar with each and every instrument and their modes of operation before use. A brief introduction of some instruments used in energy audit are listed in the following section.

### 3.12.1 ELECTRICAL MEASURING INSTRUMENTS

These are devices used for taking measurements of parameters such as kVA, kW, PF, Hertz, kVAr, Amps and Volts, and also harmonics. An **ammeter** is used for measuring the electric current denoted by symbol amperes (A). A **voltmeter** is used to measure electrical potential differences across any two points in a circuit. Analog voltmeter uses pointer which moves across a scale in proportion to the voltage of the circuit, whereas digital voltmeters have a numerical display and digital converter.

A **clamp meter** is an electrical meter with an AC current clamp-on ammeter or tong tester. A clamp meter measures the vector sum of currents in conductors passing through the probe, depending on the phase relationship of the currents.

Such instruments are used live-line, i.e., on running motors, which does not need to be stopped. The handheld meters are used for instant readings, while the advanced ones give readings after a specific time interval as a printout.

### 3.12.2   ENERGY METER

An electricity meter or energy meter is a device for measuring the energy consumed by a residence, business, or an electrically powered device, the most common one being the kilowatt-hour [kWh] found in the billing units.

### 3.12.3   COMBUSTION ANALYZER

Hydrogen ($H_2$) and carbon monoxide (CO) emissions in the flue gas are due to improper/incomplete combustion. Combustibles of more than 100 ppm is released as a waste in the form of wasted fuel, formation of soot, and leading to a reduction in the heat transfer efficiency. Moreover, a high percentage of combustibles cause a detrimental effect on the environment and a possibly volatile situation. There are three common ways of on-line keep track on combustibles in flue gas namely; non-dispersive infrared absorption, wet electrochemical cell and catalytic element. Non-dispersive infrared absorption and wet electrochemical cell techniques measure CO only. However, it has in-built chemical cells to measure numerous gases like CO, $O_2$, $SO_x$, and $NO_x$.

### 3.12.4   FUEL EFFICIENCY MONITOR

Basically, this instrument measures temperature and oxygen of the flue gas. Calorific values of typically used fuels are fed into the microprocessor and finally, it estimates the combustion efficiency of the energy conversion devices.

### 3.12.5   FYRITE

Fyrite gas analyzers are fast, precise and user-friendly instruments for measuring and analyzing carbon dioxide or oxygen. These are available for either $CO_2$ or $O_2$ analysis. A hand bellow pump suck sample of the flue gas into the solution which is available inside Fyrite. The chemical reaction changes the liquid volume and discloses the amount of gas.

### 3.12.6   CONTACT THERMOMETER

It consists of thermocouples that measures temperature of hot water, flue gas, and hot air using probe into pathway of the stream. There are different probes shaped for different requirement. For example, a leaf type probe is used to measure surface temperature while round-shaped probe is used for measuring fluid temperature in the flowing stream with the same instrument.

### 3.12.7  Infrared Thermometer

Infrared (IR) thermometer is a non-contact measuring device that gives its reading on pointing out the heat source. The basic and most affordable IR thermometer has a measurement range from 0 to approximately 600°F with an accuracy of ±3.5°F. It is used for measuring hot spots in furnaces, intricated surface temperatures, motor surface temperature, etc.

### 3.12.8  Pitot Tube and Manometer

Pitot tube is a flow measurement instrument which is used to measure the fluid flow velocity. It is extensively utilized to estimate the aircraft airspeed, water speed of a boat, and also to measure liquid, air and gas velocities in industrial applications. Pitot tube is employed to measure the local velocity at a given point in the flow stream and not the average velocity in the conduit or pipe.

### 3.12.9  Water Flow Meter

Non-contact water flow measuring devices are based on Doppler effect/ultrasonic principle. Ultrasonic water meter has an ultrasonic transducer for sending ultrasonic sound waves through the fluid to explore the velocity and convert the velocity into fluid volume ($m^3$ or liters). Transmitter and receiver of the instrument are positioned on opposite sides of the pipe/conduit.

### 3.12.10  Speed Measurements

Speed measurements are critical and a challenging task in audit exercise as it may change with frequency, belt slip, and loading. A tachometer is an instrument measuring the rotation speed of a shaft or disk, as in a motor or other machine in revolutions per minute (RPM). Contact type tachometer can be used wherever direct access is possible. More sophisticated and safer ones are non-contact instruments such as stroboscopes.

### 3.12.11  Leak Detectors

Ultrasonic instruments are used to detect leaks of compressed air and other gases which are usually not able to be detected with human abilities (eyes or ear or nose).

### 3.12.12  Pyranometer

Pyranometer is a type of actinometer which is used to measure broadband solar irradiance on a planar surface and is a sensor that is designed to measure the solar radiation flux density (W/m²) from a field of view of 180 degrees. The solar radiation spectrum extends approximately from 300 to 2800 nm. Pyranometer usually cover that spectrum with a spectral sensitivity that is as "flat" as possible. The response to "beam" radiation varies with the cosine of the angle of incidence, in order that there will be a full response when the solar radiation hits the sensor perpendicularly, zero response when the sun is at the horizon, and 0.5 at 60 degrees angle of incidence. Pyranometer should have a so-called "directional response" or "cosine response" that is close to the ideal cosine characteristic.

### 3.12.13   LUX METERS

Illuminance and light distribution have a crucial influence on performance and occupational safety. The luminous power (lumen) between a light source and illuminated area is measured using the lux (lx) unit. The illuminance is precisely one lux when luminous power of one lumen (lm) homogeneously illuminates an area of one square meter. A lux measuring instrument is referred to as a "lux meter" or also as a "photometer." Photo cell senses the light output and converting it to impulses of electricity, which is said to be lux. Few lux meters have internal memory or a data logger for recording the measurements. Lux meters with data loggers are very much useful since their cosine correction of the angle of the incident light. These lux meters can store values since it has memory and software in addition to different interfaces. It helps in transferring data to a computer, where further analysis can take place.

### 3.12.14   ANEMOMETER

An anemometer is a device for wind speed measurement which is generally used in weather monitoring station. The term originates from the Greek word "anemos" which means wind. It is divided into two classes: wind speed measurement and wind pressure measurement. Since both are important hence a single device is made for both purposes.

## QUESTIONS

Q.1.   What is "Energy audit"? Explain the types of energy audit in brief.

Q.2.   Explain the managerial functions required in energy in the Management of energy.

Q.3.   Explain why managerial skills are as necessary as technical skills in the management of energy.

Q.4.   What are the various processes in the carrying of management of energy in an organization?

Q.5.   State the importance of energy policy for industries.

Q.6.   Illustrate the use of training and awareness in the management of energy programs.

Q.8.   Discuss in brief the difference between primary and detailed energy audits.

Q.9.   Explain the usefulness of knowing the energy costs.

Q.10. Explain the usefulness of benchmarking energy consumption.

Q.11. Explain the implications of part-load operation equipment of energy using examples.

Q.12. Explain the term fuel substitution. Show examples.

Q.13. Illustrate the parameters that can be measured by on-line power analyzer.

Q.14. Name the device used for measuring $CO_2$ from boilers stack.

(a) Infrared thermometer (b) Pyrite (c) Anemometer (d) Pitot tube

Q.15. Non-contact flow measurement can be carried out by

(a) Orifice meter (b) Turbine flow meter (c) Ultrasonic flow meter
(d) Magnetic flow meter

Q.16. Illustrate the Energy Conservation Act 2001 and its features briefly.

Q.17. Name any three essential provisions of the EC act, 2001, as applied for the designated consumers.

# 4 Material and Energy Balance

When a specified quantity of different materials passes through various operations, it can be termed as material balancing or specifically "conservation of mass." The energy balance can be explained by the energy quantities and it is known as energy conservation. It means if there is no accumulation of energy, input must be equal to output. This is also applicable for any process occurring in a batch operation and it is also same for all operations of continuous nature done in any particular interval of time.

The balancing is an essential aspect of energy and materials in industrial processes. Materials are the main parameter to control the processes occurring, and more specifically, it helps to control the quality and quantity of the product. Materials are first obtained during the beginning of a new process in a plant. Experiments on the plant materials take place along with testing and planning. The examination is carried out before the commission of the plant and it is followed by periodic refinement and maintenance of the controlling instruments during the production process. If there is any specific change in the process, then material balancing must be carried out again.

With the gradual increase in the cost of energy, the industries are searching opportunities to reduce the energy consumption during the various processes in the production. Balancing of energy is carried out at various stages of the process starting from the input of raw material to finished product (output).

Material and energy balancing are general approaches of conservation of material and energy used during the various process at different stages. However, sometimes it is easy and sometimes it is complicated. Working with simple systems such as single operating units will eventually help in handling complex systems. Nowadays, with the help of computers, energy and mass balancing can be done easily with limited time for highly complex system. Hence, it can be used in everyday process control for solving processing problems to obtain good quality product with minimum costs.

## 4.1 BASIC PRINCIPLES OF ENERGY AND MASS BALANCES

The nature of the operation taking place as a single unit is represented by a block diagram, which is given in Figure 4.1. The input of energy and mass must be balanced with the output of the same, respectively.

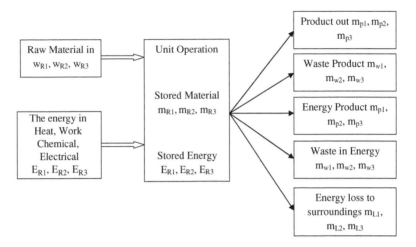

**FIGURE 4.1**   Balancing diagram of mass and energy.

According to the law of mass conservation:

Material input = Material coming out + Material remained

Raw matter = Obtained goods + Left out + Materials stored

$$\Sigma w_R = \Sigma w_P + \Sigma w_W + \Sigma w_S$$

where $\Sigma$ represents the summation,

$\Sigma w_R = \Sigma w_{R1} + \Sigma w_{R2} + \Sigma w_{R3}$ = Total raw materials
$\Sigma w_P = \Sigma w_{P1} + \Sigma w_{P2} + \Sigma w_{P3}$ = Total products
$\Sigma w_W = SM_{W1} + SM_{W2} + SM_{W3}$ = Net product waste
$\Sigma w_S = \Sigma w_{S1} + \Sigma w_{S2} + \Sigma w_{S3}$ = Total stored products

If there is no chemical changes, the mass conservation is applicable individually for every situation, so for A:

$w_A$ materials coming in = $w_A$ materials coming out + $w_A$ materials left

In case of a sugar-producing plant, the net amount of sugar fed in must be equal to the net amount of purified sugar coming out and the remaining waste left out (in the form of liqure). If there is any substantial difference in the result, then the mass balance obtained is wrong. For such case:

$$M_A = \left( w_{AP} + w_{AW} + w_{AU} \right)$$

where $w_{AU}$ is the unwanted loss which has to be obtained. Hence the balancing is:

**Raw input = Output + Wasted material + Stored material + Material lost**

where the obtained losses are unknown materials.

Similar to the conservation of mass, energy is also conserved in the processing of food. The input energy can be balanced with the output energy and stored energy in the process.

$$\text{Energy Input} = \text{Energy Output} + \text{Energy Stored}$$

$$\Sigma Q_R = \Sigma Q_P + \Sigma Q_W + \Sigma Q_L + \Sigma Q_S$$

where

$\Sigma Q_R = Q_{R1} + Q_{R2} + Q_{R3} + \ldots = \text{Net energy input}$
$\Sigma Q_P = Q_{P1} + Q_{P2} + Q_{P3} + \ldots = \text{Net energy output from products}$
$\Sigma Q_W = Q_{W1} + Q_{W2} + Q_{W3} + \ldots = \text{Net energy output from wastage}$
$\Sigma Q_L = Q_{L1} + Q_{L2} + Q_{L3} + \ldots = \text{Net energy left to the surroundings}$
$\Sigma Q_S = Q_{S1} + Q_{S2} + Q_{S3} + \ldots = \text{Net storage}$

The balancing of energy is more complex due to its interchangeability into other forms of energy, such as chemical energy to heat energy, electrical energy to mechanical energy, etc. However, it must be balanced.

## 4.2 USAGE AND THE WORKING OF THE SANKEY DIAGRAM

The Sankey diagram is used for indicating the whole energy flow. It includes the input, output and released energy (losses) in a machine or device or system, for example, boiler, heater, heat exchanger, etc. Sankey diagrams specify the different types of outputs, losses and stored energy for the energy managers to find opportunities to reduce the energy losses and to improve the overall system based on priority.

## 4.3 MATERIAL BALANCES

The three basic categories associated with mass balance are input material, output material, and stored material. For such case, material based on the category is to be considered before deciding upon the method of treatment, i.e., altogether, a net balance of mass or separate treatment for the components and based on its constituents. For example, dry solids can be taken as one kind of material. More in-depth differentiation can be done based on the composition and physical appearance. The selection of mass balance depends on the cost of raw material, different operations associated to get the finished product and the necessity of mass balance. Since the value of materials in the industry is an essential factor, the choice of selecting inexpensive material and the amount of wastage left is always taken into account.

### 4.3.1 BASIS AND UNITS

When choosing the elements that are specifically considered, a significant platform is to be decided to carry out the calculations. There can be some amount of unprocessed materials entering a system, which is divided into batches or few amount of masses entering in an hour, in a constant process. The constituent can be of a mass of something which is predominant in the process. For instance, balancing mass in a bakery is entirely dependent on the input of 100 kg of flour entering or any other material which do not change. For example, in calculation regarding combustions, where air is the main component is used to relate the nitrogen content in the compound. There are times when it is of no use to which component the materials are compared to then the amount of raw materials used in the constant process is taken into account. On deciding the matter, it is decided on what basis the calculation will be done, for example, mass or amount of concentration such as weight or molar.

### 4.3.2 NET MASS AND COMPOSITION

The balance of material is generally based on the net mass or dry solid or for any specific part.

#### Example: Constituent Balance

Skimmed milk is made after some fat reduction from the entire quantity. The contents of the milk were 90.5% water, 5.1% carbohydrate, 3.5% protein, 0.1% fat, and 0.8% ash. The real milk has 4.5% fat content, the objective is to find out the composition of the fat that was removed from the milk to make skim milk while it is considered that no losses are taking place.

100 kg of skim milk contains 0.1 kg of fat. Assuming the amount of fat that has to be removed be y kg.

$$\text{Net real fat} = (y + 0.1) \text{ kg}$$

$$\text{Net real mass} = (100 + y) \text{ kg}$$

Given, the real fat constituent is 4.5%

$$(y + 0.1)/(100 + y) = 0.045$$

where $y + 0.1 = 0.045(100 + y)$

$$y = 4.6 \text{ kg}$$

Hence, the milk composition is found to be = 4.5%, water = 90.5/104.6 = 86.5%, protein = 3.5/104.6 = 3.3%, carbohydrate = 5.1/104.6 = 4.9% and ash = 0.8%.

### 4.3.3 CONCENTRATIONS

There are many ways to express the concentration, such as in terms of weight/weight (w/w), weight/volume (w/v), molar concentration (M), and mole fraction. The weight/weight concentration is the solute weight by total weight; hence, it can be said as the fractional form of weight denoted in percentage. The weight volume concentration is the solute weight per net solution volume. The molar concentration is the solute net molecular weight represented in kg in 1 m³ of the solution. The mole fraction is a ratio of the number of moles to the net moles of the entire group of materials present in the solution. If looked closely that for process engineering, it is represented by kg moles, but in this case, the term mole means a mass of the material equal to its molecular weight in kg. For this chapter, the weight percentage is represented (w/w) unless mentioned otherwise.

### Example: Concentrations

A solution is made by adding 20 kg of salt to 100 kg of water, for making the density of liquid 1323 kg/m³. Find out the salt concentration of (a) fraction of weight, (b) fraction of weight/volume, (c) mole fraction, and (d) molar concentration.

a. Fraction of weight:

$$20/(100 + 20) = 0.167: \% \text{ weight/weight} = 16.7\%$$

b. Fraction of weight/volume:
   The given density is 1323 kg/m³, hence we can say 1 m³ of solution weighs 1323 kg, but 1323 kg of salt solution contains

$$(20 \times 1323 \text{ kg of salt})/(100 + 20) = 220.5 \text{ kg salt/m}^3$$

1 m³ contains 220.5 kg salt.
Fraction of weight/volume = 220.5/1000 = 0.2205
And so weight/volume = 22.1%
c. Moles of water = 100/18 = 5.56

$$\text{Moles of salt} = 20/58.5 = 0.34$$

$$\text{Mole fraction of salt} = 0.34/(5.56 + 0.34) = 0.058$$

d. The molar concentration (M) is 220.5/58.5 = 3.77 moles in m³

Here, the mole fraction can also be written as the number of dominant moles of water, which is near about 0.34/5.56 = 0.061. On dilution of the solution, the approximation increases, and for solutions of dilute nature, the solute mole fraction is very much similar to the moles of solute/moles of solvent.

All these methods are possible in solid/liquid mixtures, but for reliability, it is generally represented as simple weight fractions.

For a gas, the measurement of concentration is calculated in weight concentrations/unit volume, or as partial pressures. The relation can be found from the following law:

$$pV = nRT$$

where:
  p is the pressure,
  V is the volume,
  n is the number of moles,
  T is the temperature, and
  R is the gas constant which is equal to 0.08206 $m^3$ atm/mole K.

Then the molar concentration of a gas is

$$n/V = p/RT$$

Hence, the concentration weight is calculated as $nM/V$, where M is the gas molecular weight. The pressure in the SI unit is $N/m^2$, which is called the Pascal (Pa). For convenience, standard atmospheres (atm) are often used as pressure units; the conversion is 1 atm = $1.013 \times 10^5$ Pa, or very nearly 1 atm = 100 kPa.

## Example: Air Composition

In a given sample of air where 77% by weight of nitrogen and 23% by weight of oxygen calculate:

  a. The average molecular weight of air,
  b. Oxygen mole fraction, and
  c. Oxygen concentration in mole/$m^3$ and kg/$m^3$ given that net pressure is 1.5 atm, and the temperature is 25°C.

  a. Let the weight of air be 100 kg; then 77/28 moles of N and 23/32 moles of $O_2$.
     Answer:

$$\text{Net moles} = 2.75 + 0.72 = 3.47 \text{ moles.}$$

  b. Average molecular weight of air = 100/3.47 = 28.8
  c. Fraction of mole for oxygen = 0.72/(2.75 + 0.72) = 0.72/3.47 = 0.21
  d. Substituting the values in the gas equation V = 1 $m^3$, where temperature T is 25°C = 273 + 25 = 298 K, we get:

$$1.5 \times 1 = n \times 0.08206 \times 298$$

$$n = 0.061 \text{ mole/m}^3$$

$$\text{Air weight} = n \times \text{Mean molecular weight}$$

$$= 0.061 \times 28.8 = 1.76 \text{ kg/m}^3$$

Here, oxygen content is 23%, so oxygen weights $= 0.23 \times 1.76 = 0.4$ kg in 1 m³
Oxygen concentration $= 0.4$ kg/m³
or $0.4/32 = 0.013$ mole/m³
When gas is dissolved in a liquid, the mole fraction of the gas in the liquid is found out to be the first finding from the number of gas moles from the gas laws followed by taking the liquid volume and then calculating the moles from the liquid.

## Example: Gas Composition

During the process of carbonating fizzy drinks, the net amount of $CO_2$ required is similar to 3 vol of gas to 1 m³ of $H_2O$ at 0°C and atm pressure. Find (a) the fraction of mass, (b) the fraction of mole of $CO_2$ for the drink, assuming all other constituents except $CO_2$ and $H_2O$ to be ineffective.
Answer:

$$\text{Let } 1 \text{ m}^3 \text{ of } H_2O = 1000 \text{ kg}$$

$$\text{Volume of } CO_2 \text{ added} = 3 \text{ m}^3$$

$$\text{Using the equation } pV = nRT$$

$$1 \times 3 = n \times 0.08206 \times 273$$

$$n = 0.134 \text{ mole.}$$

$$\text{Molecular weight of } CO_2 = 44$$

$$\text{Weight of } CO_2 \text{ added} = 0.134 \times 44 = 5.9 \text{ kg}$$

a. Fraction of mass $CO_2$ in drink $= 5.9/(1000 + 5.9) = 5.9 \times 10^{-3}$
b. Fraction of mole $CO_2$ in drink $= 0.134/(1000/18 + 0.134) = 2.41 \times 10^{-3}$

## 4.3.4 TYPES OF PROCESS SITUATIONS

### 4.3.4.1 Continuous Processes
When a process is continuous, time is a critical parameter during balancing and is done on the basic unit of time. Hence, if a centrifuge of eternal nature for the milk separation from the skimmed milk and cream, and the material used in the centrifuge has no changes related to the composition and also the mass, then the quantity of input and output in several stages on the basis of time taken as the unity for the equal mass of any changed balanced mass can be found out. Studies show that the process takes place in a steady-state, and there is no change in the flow and matter during that entire time period.

**Example: Balancing among Equipment in a Centrifuge
of Continuous Nature for Milk**

A container ensembles 35,000 kg of net milk consisting 4% of fat that is removed
in 6 h to produce skim milk having a fat of 0.45% and cream with 45% fat, then
find the flow rate of the outputs of a centrifuge of continuous nature that performs
the removal.

Taking 1 h as the milk flow

Mass in

Answer:

$$\text{Net mass} = 35,000/6 = 5833 \text{ kg.}$$

$$\text{Fat} = 5833 \times 0.04 = 233 \text{ kg.}$$

$$\text{Water solids} - \text{Not fat} = 5600 \text{ kg.}$$

Flowing out of mass

Taking cream mass to be y kg, then net fat is 0.45y. Then of skim milk
(5833 − y) kg and its net fat content is 0.0045(5833 − y) kg.

A material balance on fat:

$$\text{Fat in} = \text{Fat out}$$

$$5833 \times 0.04 = 0.0045(5833 - y) + 0.45y \text{ and so } y = 465 \text{ kg.}$$

Cream flow is 465 kg/h and skim milk (5833 − 465) = 5368 kg/h

For continuous processes, the time factor should be handled carefully since the
processes work only for a segment of the entire time of the factory. It is divided
into three time zones, such as starting, steady-state process, and closing down,
and the study of the material which is to be balanced is an essential factor here.
The time interval should be long during such cases since there may be slight variations periodically.

In some cases, chemical reactions take place and balancing of the material
needs to be customized according to those reactions. A chemical change may
takes place, such as destroying bacteria during the healing process. The net mass
in this case for a system will have no change, although the inside areas can change,
such as in sugar browning there is reduction in sugar components, whereas, it
leads to increasing amount of browning compounds.

### 4.3.4.2 Blending

Another type of situation is blending. In blending, various materials are mixed in
different proportions to get a desired product. For example, in producing mineral
water, minerals are added to make it more appropriate for human consumption,
where it is solved using direct balancing of material.

### 4.3.4.3 Drying

During the balancing of the materials, equations can be developed for the individual
process taking place and the entire process as a whole. For cases where there is a

group of material having a constant ratio, the equations are generally composed for groups rather than individual processes. For example, during vegetable drying, the components such as minerals, proteins are classified as dry solids and the whole drying process consists of dry matter and moisture content for balancing of material.

## 4.4 ENERGY BALANCES

There are different forms of energy such as chemical energy, heat energy, potential energy, kinetic energy, etc., and they are subjected to interconversion; hence, it is complicated to balance energy individually based on its components. However, for some cases, specific balancing is possible due to limitations and situations. In few cases, all types of energy cannot be taken into account, for example, if water flows through a pipe then heat generation due to friction of flow is not essential in energy balance.

In practical situations, the balancing helps in focusing on certain essential situations, such as for a heat balance, it can be used for describing and figuring out the cost and quality associated with the process. During unknown situations of energy loss during balancing, it is best to ignore such balancing factors. After calculation, the significant aspects are to be taken into consideration, and the remaining as a whole is ignored. The minor losses are to be identified and eliminated carefully; otherwise, it may lead to an error in the balancing.

The balancing of energy is also calculated for external cases, such as the amount of raw material processed. For example, in the food industry, the direct energy associated is the cost of fuel and electricity, whereas indirect energy is related to machines, packaging, etc.

In the SI unit, energy is represented by joule. Kilocalories are another unit and British thermal units (Btu) in work related to heat balancing.

The application of energy transfer in this chapter is for fluid flow and heat transfer.

### 4.4.1 HEAT BALANCES

The most widely used form of energy balance is heat, and its balancing is done based on heating and cooling. Here, the total enthalpy undergoes conservation followed by the conservation of mass; hence, enthalpy balancing is done for the different types of equipment or processes, etc., and it is assumed that there is no interconversion of heat to another form.

Enthalpy (H) is obtained based on some reference frames, hence the values and the parameters are relative to that reference frame. In addition, energy balance is just taking proper and correct quantities of the materials in terms of temperature, and specific heat, etc. Figure 4.2 shows the balance of heat.

The heat is either accepted or given out in small amounts in reactions process but in comparison with the amount of matter entering in the processing of food such as sensible heat and the latent heat. The latent heat is used for changing the state of the matter at a constant temperature. The sensible heat is the heat, which is required to increase or decrease the temperature from the material.

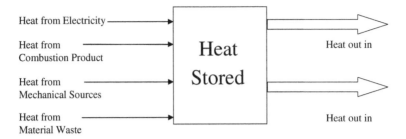

**FIGURE 4.2**   Heat to surroundings.

The specific heat can be expressed by

$$\text{Specific heat} = m \times c \times \Delta T \quad J/kg \cdot K \, (\text{unit})$$

where:
   m is the mass,
   c is the specific heat, and
   $\Delta T$ is the change in temperature.

The latent heat (L) can be expressed as

$$L = Q/m \quad kJ/kg \, (\text{unit})$$

where:
   Q is the amount of heat and
   m is the mass.

With the help of these equations, energy balance can be done easily and significantly. The calculations associated behind these balancings are straightforward and help on a quantitative basis later on for designing the process and the equipment. It can help to design or evaluate the desired equipment and suitable process.

### Example: Dryer Heat Balance

Textile drier consumes 4 m³/h of natural gas having a calorific value of 800 kJ/ mole. If the thorough output of the drier is 60 kg, moisture is dried from 55%–10%, then find the net thermal efficiency for the drier considering only the latent heat of evaporation.
   60 kg of wet cloth contains

$$60 \times 0.55 \text{ kg water} = 33 \text{ kg moisture,}$$

$$60 \times (1 - 0.55) = 27 \text{ kg bone dry cloth.}$$

Final moisture 10%, hence the moisture is 27/9 = 3 kg.

Removal of moisture/h = 33 − 3 = 30 kg/h

Latent heat of evaporation = 2257 kJ/K

Heat for supply = 30 × 2257 = 6.8 × 10⁴kJ/h

Assuming the natural gas to be at standard temperature and pressure (S.T.P) where 1 mole occupies 22.4 l.

Flow rate of natural gas = 4 m³/h = (4 × 1000)/22.4 = 179 moles/h

Available heat for combustion = 179 × 800 = 14.3 × 10⁴kJ/h

The thermal efficiency of dryer = Heat needed/ Heat used

$$= 6.8 \times 10^4/14.3 \times 10^4 = 48\%$$

For better evaluation of efficiency, it is essential to take into consideration the heat associated with the dry cloth, moisture content and other factors such as temperature humidity, etc., which will be added to the natural gas. Here as the latent heat of evaporation is the main and important factor to be considered; hence, it is easier and fast to give essential information by the energy balance.

Also, energy balance can be done for different heat processing operations related to thermal energy or any other form of energy.

### Example: Autoclave Heat Balance in Canning

1000 cans of pea soup are present in an autoclave; heating is done up to a temperature of 100°C.

To cool it to 40°C before leaving the autoclave, find the amount of cold water needed for input at 15°C and leaving at 35°C.

The specific heats mentioned are 4.1 kJ/kg °C for pea soup and 0.50 kJ/kg °C for can. Each can weight is 60 g containing 0.45 kg of soup. Assuming that walls of the autoclave having temperature above 40°C is 1.6 × 104 kJ and no loss through the walls.

Let w = required weight of cold water; 40°C = datum temperature, the temperature at which cans are leaving the autoclave (Table 4.1).

---

## TABLE 4.1
## Balance of Energy in the Form of Heat in the Process of Cooling; The Datum Line is 40°C

| The Entry of Heat (kJ) | | Left Out of Heat (kJ) | |
|---|---|---|---|
| Can heat | 1800 | Can heat | 0 |
| Can contents heat | 11,000 | Can contents heat | 0 |
| Autoclave wall heat | 16,000 | Autoclave wall heat | 0 |
| Water heat | −104.6 w | Water heat | −20.9 w |
| Net heat input | 127.800−104.6 w | Net heat output | 20.9 w |

### 4.4.2  ENTRY OF HEAT

Heat in cans = Can weight $\times$ Specific heat $\times$ Datum temperature

$$= 1000 \times 0.06 \times 0.50 \times (100 - 40) \text{ kJ} = 1.8 \times 103 \text{ kJ}$$

Content of heat in can = Soup weight $\times$ Specific heat $\times$ Datum temperature

$$= 1000 \times 0.45 \times 4.1 \times (100 - 40) = 1.1 \times 105 \text{ kJ}$$

Heat in water = Weight of water $\times$ Specific heat $\times$ Datum temperature

$$= w \times 4.186 \times (15 - 40)$$

$$= -104.6w \text{ kJ}$$

### 4.4.3  LEFT OUT HEAT

Can heat $= 1000 \times 0.06 \times 0.50 \times (40 - 40)$ (cans leave at datum temperature) $= 0$

Can content heat $= 1000 \times 0.45 \times 4.1 \times (40 - 40) = 0$

Heat from water $= w \times 4.186 \times (35 - 40) = -20.9w$

Net heat input = Net heat output

$$127{,}800 - 104.6w = -20.9w$$

$$w = 1527 \text{ kg}$$

Cold water $= 1527$ kg

### 4.4.4  TYPES OF ENERGY

The source of power for the motor is electricity and it can be generated from steam power plant or hydropower. The electricity supplied is calculated using a wattmeter and power consumed by the electrical drive can be estimated. Due to heating, motors have also different types of losses such as friction loss and windage. Usually, the efficiency of motor is shown by the manufacturer and it can be calculated by the energy ratio input to the output .

If energy balance is done by the considering movement of fluid flow in pumps, material handling occurs from one place to another place and food ingredient mixing is done with the help of mixers. The flow condition is typically found out in the form of conversion of one to another and in the form of energy conservation such as friction losses, heat losses during combustion, energy released during chemical reaction, etc. It can be calculated using calorific value of fuel and its combustion rate. Generally, energy is released in the form of heat and is obtained from various sources.

### Example: Refrigeration Load

10,000 loaves of bread are frozen, having weight each of 0.75 kg from 18°C to a temperature of −18°C. The process is to be carried out in air-blast freezing. It was seen that the motor's rating of the fans was 80% of the net horsepower, but in actual, it is working at 90% of the rating. If 1 ton of refrigeration is 3.52 kW, find out the highest load of the refrigerator freezing, taking the following assumptions: (a) The fan and the motors are insulated in the tunnel for freezing and (b) the fans are in the tunnel and not the motors. The rate of heat loss from the tunnel was found to be 6.3 kW.

Rate of removal of heat-related to freezing of bread (maximum) = 104 kW
Rate of horsepower of a fan = 80

Now 0.746 kW = 1 hp and working of the motor at 90% of rated value,

And so (fan + motor) power = (80 × 0.9) × 0.746 = 53.7 kW

a. With motors + Fans in tunnel
   Load-related heat of fans + Motors = 53.7 kW
   Load of heat from ambient = 6.3 kW
   Load due to net heat = (104 + 53.7 + 6.3) kW = 164 kW
   = 46 tons of refrigeration
b. Motors external, the motor inefficiency = (1 − 0.86) not putting a load in the refrigeration

$$Total\ heat\ load = \left(104 + \left[0.86 \times 53.7\right] + 6.3\right)$$

$$= 156\ kW$$

$$= 44.5\ tons\ of\ refrigeration$$

There are cases where material and energy balance are both combined due to same stoichiometric composition.

## 4.4.5  Summary

1. Material and the energy balance can be used to calculate quantitatively by the help of amount of material fed to a process and types of process.
2. Material and energy balances include the input content = finished product content + wastage of contents during the process + storage content.
3. Time balance must be considered during the continuous process.

4. Energy balance comprises all types of energy such as kinetic energy, potential energy, chemical energy, and heat energy. Energy balance is the sum of all kind of energy which can be conserved.

5. In enthalpy balance, heat is the most important factor in different types of operations and conditions.

   The main objective of material and energy balance is to measure the input of material/energy, its conversion efficiency, final output and different types of losses. Energy and material balance can be used as powerful tools to analyze the equipment performance and processing operations applied. With the help of analysis report, the energy manager can suggest the required necessary improvement to the management and it can save the energy and material finally in reducing the wastage.

## 4.5  PREPARATION OF PROCESS FLOW CHART

In energy and material balance, identification of proper operation or process is very important. An example of flow chart is given in Figure 4.3. The flow chart is a schematic view of complete production process which comprises different types of inputs, different operations, conversion techniques and different outputs. The process flow chart comprises the stepwise process at every stage such as input, operations, output and losses or wastage. The flow of the process is to be done stepwise for better identification of each situation, as shown in Figure 4.3.

**Input** includes different types of raw materials such as electricity, material, fluid, etc.

**Process** includes the number of process/operations required to convert the input material into final product, by product, and intermediate process. The various operating parameters of the processes should also be mentioned. The rate of flow of steams should be appropriately given using units like m³/h or kg/h. For a process undergoing as a batch, time must be included as a parameter.

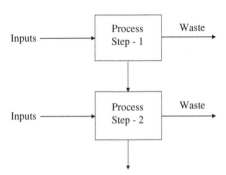

**FIGURE 4.3**   Flow chart of the process.

**Wastes**/by-products may contain various types of matter and materials. For every single step, the balance of material and energy should be done for the individual processes, and the entire process of the plant.

**The output** is the final product obtained in the plant.

## 4.6   FACILITY AS AN ENERGY SYSTEM

Many systems of energy are present to provide a secondary form of energy for a manufacturing or process plant. An energy system is represented in Figure 4.4. Even though the intake forms are different and are taken into account, the final output is obtained as a form of heat, having quite low/high temperature. Primary energy sources (coal, oil, electricity, etc.) are used here to form a secondary energy (warm water, steam, high pressure air, high/low temperature, etc.)

The use of energy is divided into different stages, as illustrated:

- Electrical energy bought as HT and finally, obtained as LT for the final use.
- Separate electricity is generated from DG sets or captive power plants.
- Fuels of several types are converted into electricity.
- The boiler produces stream for heating and for drying purposes.
- The cooling tower and supply of cold water are used for cooling purposes.
- Air compressor and air compressor supply are used for compressed air needs.

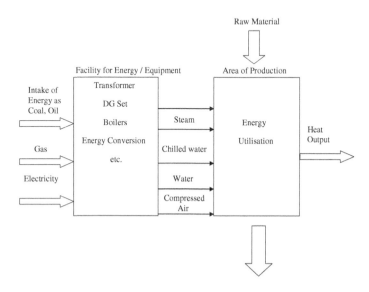

**FIGURE 4.4**   System for the energy for plant.

All system of energy is categorized into three types such as generating, distributing, and utilizing for a proper approach and its analysis. Few such examples are given as follows:

### 4.6.1  BOILER SYSTEM

The boiler and its accessories and mounting are used for the analysis of energy. A diagram can be drawn as in Figure 4.5, to do the material and energy balance. The diagram also includes the various subsystems such as the feed of water system, economizer, air preheater, superheater, fuel handling system, etc., and the supply of steam, etc.

### 4.6.2  COOLING TOWER

Cooling towers or chillers are very common in any industry. A complete diagram of the cooling tower system is shown in Figure 4.6 for the analysis and audit. Every significant usage of cold water is to be shown in the diagram.

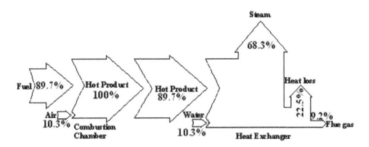

**FIGURE 4.5**   Flow energy diagram for a boiler plant. (From Saidur, R. et al., *Energy Policy*, 38, 2188–2197, 2010.)

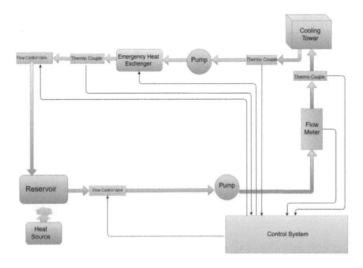

**FIGURE 4.6**   Cooling tower water system. (From Hossain, A. et al., *Energy Procedia*, 160, 566–573, 2019.)

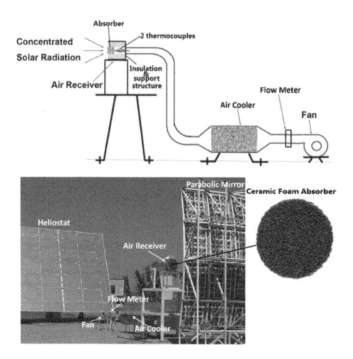

**FIGURE 4.7** Types of equipment related to air systems. (From Li, Q. et al., *Appl. Energy*, 178, 281–293, 2016.)

### 4.6.3 System of Compressed Air

Compressed air is widely used in many industries. It is very safe and also useful. A general system of the compressed air is shown in Figure 4.7. Analysis of energy and the practices measured should be illustrated.

## 4.7 PROCEDURES IN CARRYING OUT BALANCING IN ENERGY AND MASS

The main reason for carrying material and energy balance is for identification of the unknown losses of energy and material during different process and operations. By proper observations, it can help to improve the plant/energy efficiency and can reduce the product cost. A proper material and energy balance is useful to reduce the utilization of raw material also. It can also help to reduce the environment pollution.

### 4.7.1 Guidelines for Material and Energy Balance

- For a complex production system, material and energy balance is a very useful tool.
- Instead of whole system, splitting into small systems is easy and simple for energy and material balance. The formation of process flow chart may be more suitable.

- The material and the energy balance flow should be selected such that there should be very less number of system, which is to be taken into consideration for the input and the output.
- Streams for recycling is to be chosen always belonging inside the envelope.
- The calculation can increase the involvement of factors such as time and production links.
- Take into account the entire batch for a batch operation.
- Take into account of the consumption of energy during starting and cleaning.
- Calculate the volumes of gas at STP.
- The mean of the process should be taken for an extended period.
- Point out the losses and the emissions (M&E) in part operations, if possible to ignore any shutdown effect
- Denote the quality of energy for each stream wherever possible.
- During energy and mass balance for a short period, the precision of the data is to be maintained to avoid the error.

The (M&E) balancing is done along with the guidelines shown previously, and they are being created at different stages:

1. Overall material and energy balance: It includes the total input and output for the entire plant.
2. Material and energy balance section/department/center basis: This is done based on each department or section in the plant and is useful for finding out the areas of modification/improvement to avoid the losses.
3. Material and energy balance equipment/device basis: Material and energy balance, for the primary devices, would help in the assessment of the performance of specific devices/equipments. It would help in finding out the losses in the energy and losses which can be avoided.

### 4.7.2   PROCEDURE FOR ENERGY AND MASS BALANCE CALCULATION

The mass and energy balance is a process to check whether the condition of energy conservation is valid for any process.

It is a critical practice and is a necessary procedure for finding out the energy and mass conditions in general. For correct use, the process illustrated below should be followed:

- Identification of problem which is able to study.
- Define a boundary for the whole system or the subsystem for the analysis, and the energy and mass balance must be done by following the boundary condition.

- The boundary condition must be selected so that:
  a. Taking all related flows must cross it and the other flows are to be situated inside the boundary.
  b. Easy and accurate measurements at the boundary should be possible.
- Choose an appropriate product testing period based on process/operation.
- Perform proper measurement.
- Determine the mass flow and energy.
- Check the mass and energy balance, if there is any error then repeat it again.
- The endothermic or exothermic processes taking place are to be taken into account during the energy balance.

**Example/Formula**

Supply of energy for combustion: Q = Burned fuel × Net value in calories

**Example 1: Heat Balance in a Boiler**

Balancing of heat is a method to balance the net heat entering and the net heat leaving the system in various forms, such as in this case, for a boiler.

## QUESTIONS

Q.1. Draw a diagram for input and output of the process, pointing out the energy inputs.

Q.2. Explain the reason for the balancing of energy and materials.

Q.3. Explain the usefulness of the Sankey diagram during the analysis of energy.

Q.4. Create a flow chart of a product for the manufacturing process.

Q.5. Illustrate the rules and regulations for energy and material balancing.

Q.6. Balancing of material is done based on
   (a) Mass, (b) Volume, (c) Concentration, and (d) Temperature.

Q.7. Continuous baking of biscuits is to be done in an oven. Moisture content entering is 25%. The moisture content at the exit is 1%. 2 t/h is produced daily being at a dry state. Perform a balance of materials and also find out the moisture amount removed per hour.

Q.8. The loading of materials in the furnace is 5 T/h. Losses in scales 2% find the material output.

Q.9. Inlet and outlet temperature for a heat exchanger is 28°C and 33°C. The circulation of cold water is at 200 L/h. The fluid entering the heat exchanger for processing is 60°C and leaving at 45°C. Calculate the fluid flow rate for the process. (Cp for the process of fluid = 0.95)

Q.10. Boiler steam output is measured using feed water. The reading of the tank from morning 8 to evening 8 was 600 m³. A continuous blowdown was provided at the boiler, having a feeding rate of 1% for that hour. Calculate the mean steam originally obtained per hour.

Q.11. Given below are the requirement for cold water in an industry for a process:

Heat exchanger 1: 300 m³/h at 3 kg/cm²

Heat exchanger 2: 150 m³/h at 2.5 kg/cm²

Heat exchanger 3: 200 m³/h at 1 kg/cm²

Calculate the remaining cold water required per hour in the industry (the HE are in parallel).

Q.12. The dryer condensates at a rate of 80 kg/h, and the calculated flash stream is 12 kg/h. Calculate the original consumption of steam for the dryer.

# 5  Energy Action Planning

Energy action plan (EAP) is an outline used by governments and large-scale company to furnish their current energy consumption and frame policies to reduce energy consumption. Energy efficiency is enormously significant to all energy intensive institutions. There are four key necessities for an effective energy management program as shown in Figure 5.1. Any efficacious energy management program within a company desires the entire support of top management. Therefore, the support of top management is the main prerequisite for success. The energy efficiency should have similar priority like raw materials, manpower, production, and sales in any organization. The other significant necessities are; a sound strategy plan, an efficacious monitoring system and substantial technical ability for evaluating and implementing energy saving choices.

## 5.1  ENERGY MANAGEMENT SYSTEM

Companies expect for a financial return from effective and proper energy management to improve their performance of energy. The success is based on routine appraisal of energy performance followed by latest planning and implementing action plans to reform energy efficiency. Therefore, a perfect energy management system is a necessary condition for pointing out and execute energy conservation remedy, maintain the tempo and for consistent effective improvements.

### 5.1.1  SUPPORT AND COMMITMENT OF THE TOP MANAGEMENT

It is the duty of the top management to make the commitments for allocating manpower and funding for making the improvement of energy continuous. There should be commitment from apex management body for assigning human resources and financial grant to attain constant developments. Furthermore, organizations need to appoint energy manager, build a committed team, and institute an energy policy for establishing a successful energy management program.

#### 5.1.1.1  Energy Manager Appointment
The duties of an energy manager are; to set the deadlines to achieve the goals projected by the top management, monitor the progress of energy conservation activity, and promote the energy efficiency program. An organization accomplish set target of energy efficiency by forming energy performance as a main objective with the help of Energy manager. Energy manager may not be always technically genius in the energy conservation process. In contrast, an effective energy manager has an understanding of energy efficiency and helps an organization to achieve its goals environmentally and financially. The role of the energy manager depends on the size of the organization, which can be full time or an additional assigned work.

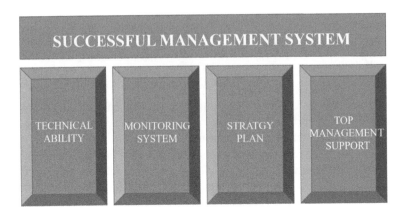

FIGURE 5.1   The main four pillars of the management of energy.

*Energy Manager Location*: The function of energy management either inherent in one "energy manager or coordinator" or allocated among the middle level managers, which is generally positioned between the senior managerial level and those who handle the energy end uses in the organization. The exact location of energy management function depends on the structure of an organization.

*Energy Manager*: Responsibilities and Duties to be Assigned Under The Energy Conservation Act, 2001 (Refer Chapter 3).

### 5.1.1.2   Dedicated Energy Team Formation

The functions of energy team are performing energy management activities across various units of the organization and make sure the amalgamation of best practices. Every day the energy use is affected by the decisions made by various employees in the organizations. Developing a team for energy management will make the management more compact for better usage of energy in the company. Energy team computes and monitors energy performance, and liaises with top management, staffs and other shareholders, other than planning and implementing particular reforms. The energy team size relies with the organization size. The energy team can have representation from each operational unit which considerably influences energy uses, for example, engineering, operations and maintenance, building/facilities management, purchasing, environmental health and safety, contractors and suppliers, utilities, etc. The constitution of the energy team varies from organization to organization, depending on the current management structure, the kind and amount of energy used and other organization-specific factors.

### 5.1.1.3   Establishment of Energy Policy

The energy policy gives the foundation to the energy management team to set the energy conservation goals and performance target in the operations by tapping the values of energy management in the working culture of the company. Energy Policy offers the strong basis for efficacious energy management implementation in the organization. It legalizes the support of top management and expresses the convictions of organization to energy efficiency for employees, shareholders, the community, etc. An officially announced energy policy acts as:

- A clear-cut statement of organization towards energy conservation and environmental protection.
- A valid document that acts as a guide regarding the practice of energy management and making consistency.

It acts as a supporting document for the company's ideas and goals toward energy management and its action plans and its declarations to abide by those plans.

**Typical Format of Energy Policy**

- Announcement of top management's convictions, and participation of senior-level and mid-level management in energy management.
- A written statement illustrating the policies.
- Official statement of company's objectives regarding short-and long-term goals.

### 5.1.2 PERFORMANCE OF ENERGY ASSESSMENT

Prior knowledge of energy usage in the organization significantly, helps in the identification of possibilities of improvement in the performance of energy utilities that increase the financial gain for the company. The energy performance assessment is a cyclic process of finding out energy usage in various units/department/facilities of the organization and founding a baseline for comparing future outcomes of energy efficiency attempts. The main features of this process are: collection of data and management, baseline establishment, benchmarking, analysis, evaluation, and conduct technical assessment and audits.

#### 5.1.2.1 Collection of Data and Management

**Collecting and Tracking Data–Collecting Information Related to Usage of Energy and Documenting Over Time.**

Calculating the performance of energy needs proper information about how, when, and where to use the energy. The collection and keeping track on this information is essential to set up baselines and management of energy usages. The process is illustrated below:

1. **Data Collection**

   There should be completed and corrected energy usage data as these will be used for further investigation and fixing the target. Consider the following steps during collection of energy use data:

   *Determine the appropriate level of detail*—The quality and purview of data collection varies from organization to organization. Some experts prefer to collect data based on individual processes from submeters whereas others choose utility bill only.

*Account for all energy sources*—Prepare a detail list of all kind of energy such as electricity, gas, steam, waste fuels, either bought or produced onsite in terms of cost basis.

*Document all energy uses*—Gather energy bills, meter readings, and other energy use data for all known sources. Energy data must be easily accessible. It can be kept in accounts department or centrally or all facility or can be received from appropriate utilities or energy service providers. Collect a recent at least two years (monthly or more frequent interval) data.

*Collect facility and operational data*—Normalization and benchmarking are to be done with the collected data; therefore, it is also essential to accumulate other data related to non-energy part from all facilities and operations (operating hours, production, building size, etc.) for future reference.

2. **Tracking of Data**

Ordinary worksheet to comprehensive databases or modern IOT systems can be used for performance tracking. The following action should be taken for proper development of tracking system for the company:

*Scope*—Tracking system is designed, based on large domain of scope and level of information which is used for tracking and frequent data collection.

*Maintenance*—Tracking system should be user friendly, easy to update and of low maintenance.

*Reporting and communicating*—The tracking systems is to be used for disseminating the energy performance results to the other units/department of the organization and inspire for changes. Develop easily understandable formats of energy performance facts across the organization. Easy reporting is expected from a good tracking system.

*Actions*

– Accumulate data of individual fuel consumption at each and every unit/building/facility,

– Accumulate data from submeters, if available,

– Apply recent and regular interval data,

– Employ advance tracking systems to generate quarterly and annual energy performance reports, and

– Utilize tracking systems to permit facilities to compare their performance with their peers.

3. **Normalize Data**

Energy consumption in utilities varies significantly mainly because of energy efficiency of the equipment and operations. However, there are various factors other than energy efficiency such as weather or certain operating characteristics which affect it partly. Normalizing is the procedure of eliminating the influence of numerous factors on energy usage. Hence, energy consumption performance of various facilities and operations can be compared. Normalization factors are needed to normalize the data.

Find the main factors needed for the comparison. These factors are based on type of the organization.

For industrial areas, the normalization factor consists of:

- Inputs
- Product type
- Output
- Production processes

For an institutional and commercial area, the normalization factor generally consists of:

- Weather/Climate zone
- Utility/Facility size
- Fuel choice
- Energy cost
- Past weather data
- Number of operation hours
- Occupancy status
- Other distinguish characteristics

### 5.1.2.2   Baseline Establishment

**Create a Baseline—Set the starting point from where the measurement is to be done.** Calculating the performance of energy at a particular time sets a baseline. It gives the initial point for fixing goals and estimating future endeavors for improving overall performance of the organization. Baselines must be found at all stages which suits to organization. The key sequences by utilizing collected data:

**Create base year**—Create a base year or an average of many past years. Utilize the major relevant and complete sets of available data.

**Identify metrics**—Choose effective and appropriate measurement units for conveying performance of energy of the organization. For example: kCal/kWh, kCal/ton, total energy cost/ton.

**Publish results**—Declare performance baselines to various facilities/units/department, managers, and other key stakeholders of organization.

### 5.1.2.3   Benchmark

**Energy performance comparison of department/unit/facilities to each other, mates, and competitors. Make preference list of facilities which need focus for further upgradation over time.** Benchmarking authorize for comparing the energy performance of the various identical facilities or a set performance level. This process is useful in the energy management as it helps in figuring out how much improvement or the relative increase in the performance of the energy occurs over time. It helps in recognizing top energy management exercises. There are many ways of benchmarking; performance of any organization can be benchmarked in the following manner:

**Performance in Past**—Comparing the previous performance with the current performance.

**Similar Industry Average**—On the basis of established performance metric. For example: the accredited average performance of an identical group.

**Best in Class**—Best among similar group of industry instead of average one.

**Best Practices**—A proper comparison on a qualitative basis on some developed methods is the best way to practice in the industry.

The significant steps in benchmarking are:

- Set the level to be benchmarked
- Create a metric system
- Compare the results
- Keep a record of the performance over time

### 5.1.2.4   Analysis and Evaluation

**Data Analysis—Figure out the latest trends and patterns in energy usage.**
Study the data to find out the trends in energy usages for understanding more effective component which influence energy performance and pinpoint the measure for lowering energy consumption. Estimation of energy performance supports in:

- Classifying present energy usages from the type of fuel, facility/operating division, product line, etc.
- Finding top functionable facilities for identifying and reusing the best practices.
- Rectify problems in worse functioning facilities and give priority in prompt reform.
- Recognize the input of energy expenses to operating costs.
- Create perspectives and references based on the past data for futuristic measures and resolutions.
- Set reference points for computing the performance for award.

Energy performance data can be analyzed by quantitative and qualitative ways as per specific need of organizations:

*Quantitative Appraisals*
- Prepare energy use profiles to find the top and bottom of energy consumption for correlating with operations or main events.
- Compare identical facilities in the industry with their energy use and performance statistics.
- Recognize areas of high-cost energy use to quantify economic impacts.
- Recognize data gaps to identify zones where further information is required.

*Qualitative Appraisals*
- Organize in-house audits or surveys/interviews for collecting information/opinion from co-workers and system-specific information (HVAC, lighting, refrigeration, etc.).
- Examine the institutional policies and operation procedures to trace impact on energy usages.

### 5.1.2.5   Conduct Technical Assessments & Audits

**To Explore the Improvement Possibility Appraise the Operational Performance of Facility Systems and Devices.**

Understanding the baseline energy use of establishment and the relative performance of complete groups are important fragments of the information needed. Periodical appraisal of the functionality of equipment, processes, and systems support in finding improvement possibilities. Energy audit is the detailed examination done by energy auditors/engineers. It assesses the actual performance of systems and equipment of any facility in contrast to designed performance level or best accessible technology. The difference between them is estimated energy saving potential. The key steps involving the audit and technical assessment are:

**Create a team for audit**—The audit team should have an expert of energy systems, process and equipment, including facility engineers, system specialists, and other support, etc. Outside support may also take for objective viewpoint or particular expertise.

**Planning and developing a strategy for the audit**—Pinpoint and prioritize systems for assessment, allocate team members to tasks, and fix the deadlines for completion the work. Apply benchmarking outcomes to find weak performing facilities and should be intended for assessment.

**Prepare energy audit report**—Produce a summary of concrete steps to lessen energy usages based on the energy audit outcomes. Report should endorse the possible actions extending from small alteration in operation to equipment/spare replacement. Determine the necessary requirement of resource for accomplishing actions and also be included in the report.

### 5.1.3   GOALS SETTING IN AN ENERGY MANAGEMENT ORGANIZATION

For achieving organizational goal, performance goals need to be set up to accomplish all the activities related to management and for proper growth and development of organization. To clarify measurable goals is a critical task for achieving desirable results, effectual strategies and to glean financial gains. Specified goals are the key factors to track, measure the progress of work and in decision making process. The motivation of staff can be maintained by posting and communicating goals in order to fulfill the organizational goal. The role of energy manager becomes important, while they develop goals with the energy team.

Once goals are set, then to work for an energy manager becomes easy like improvement and growth for entire organizational structure for effectual energy management program (EMP), EMP's success measurement, to help the EMP team in progress of the project, create purpose and keep motivating his staff members, express his commitment towards less degradation of environment and activity upgradation schedule.

Determining scope and estimation of potential of improvement are very necessary for the development of performance goals.

The scope incorporates different levels of the organization and different time duration to accomplish tasks related to any specific goal.

The different organizational levels are (i) organization wide level, (ii) facility level, and (iii) process or equipment level. At organization wide level, big picture of goals can be seen for the overall development and growth of an organization. A framework of communication for the audience inside and outside is described for successful energy project. The facility level describes the goals which incorporate the specific facility in order to get best results. The process or equipment level set goals for unique process lines and equipment, while considering energy used in a particular area.

As per the time duration, there are basically two types of goals (a) short term goals and (b) long term goals. Reporting and tracking of EMP progress are considered in annual goals which are termed as short term goals, on the other hand long term goals depend upon the organization usually affected by internal rate of return strategy horizon of planning and guidelines, and commitment for protection of environment.

Estimation of potential for improvement is also an important aspect of EMP. According to the estimation of potential of organization and economic weight age of the resources needed, the management would be able to set goals. The potential determination generally depends on the availability of resources, deadlines for the project, use of various energy resources, and organized style of energy program. Nationally and internationally, the major energy programs adopted the methods which incorporate some exercises like performance data review, benchmarking, evaluation of earlier projects, best adopted practices, auditing and technical assessment review, comparison of goals of the organizations at the same level playing field, and understanding of wide strategic goals of the organization.

After this long exercise, finally goals are established after potential estimation and the creation of measurable and clear goals with target deadlines. Energy performance goals (EPGs) are expected to be established at top level of management and are considered as mission for entire organization. Most of the organizations set their EPGs according to the factors that affect the organization and technical feasibility.

These factors include quantification of the goal, compared performance with best-in-class benchmark, efficiency improvement by reducing energy consumption per unit production, improving environment by reducing energy consumption or applying pollution prevention techniques, and the energy should be ready to achieve even aggressive goals.

Once goals are set, the project manager needs to create an action plan, however, to avoid any risk, the management must go through a field force analysis (FFA) which is very much helpful to overcome all the barriers. Deep insight about the process involved in EMP can be provided through FFA. There are some steps for this analysis to be completed like indication of the direction of the organizational goal, identifying barriers, identifying positive influences, estimation of the relativity of the positive and negative influences, and prioritize all the forces whose strength can be increased or decreased in order to achieve the organizational goal.

### 5.1.4  FORMULATION OF AN ACTION PLAN

Energy performance can be improved by developing a road map and without setting goals no organization can develop a road map. Implementation of energy performance measures can be done by applying a systematic process and a responsible and successful organization develops a broad action plan to create this systematic process. Energy policy is framed by bureaucrats and energy experts for a long period of time and more or less no change is made for that particular time period, on the other hand updating of action plan is necessary on regular basis annually for the reflection of achievements, for the performance improvement, and to reshuffle the priorities. As per the general or urgent requirement the action plan can be scaled up, it depends on the type, size and budget of the organization.

Some steps need to be followed for formulation of a plan.

1. Technical assessment evaluation and auditing—for the successful completion of EMP after a regular interval of time, the gap between present performance and goals must be determined through progress evaluation and technical assessment.
2. Technical steps- upgrade facilities to the desired performance from present performance in order to achieve the goals.
3. Performance targets—every department, operation and facility of the organization must be assigned some performance target to achieve goals.
4. Deadlines—to accomplish any task with already decided actions to be taken a deadline must be decided. Regular evaluation of progress in meeting should also be organized on scheduled deadlines.
5. Tracking system—all the action, usage of energy and program activities must be tracked and monitored.

Roles of individual and energy team to realize the action plan after finalizing from management become very important. Assign responsibility and fix accountability of individual involved in their respective task, action or duties related to their departments like department of environmental issues, communication, supply management purchase, procurement, human resources, capital investment, planning and operations management, etc. External roles also become significant for the successful completion of EMP. Raw material providers, vendors, service providers, consultants, shareholders, foreign investors, government latest orders relevant to EMP are the associated elements with an organization undergone to accomplish projects within a stipulated period of time. Without their synchronization with projects EMP couldn't be successful. Some organizations sometimes outsourced their projects. In such cases, mention all the standards in tender notice to evaluate the bids, so that they can be incorporated in the agreement with contractors. Cost is also a main factor which should be estimated for all items and human resource. Energy project must be designed in such a way that government approval and fund could be availed. Even though excellent management through extraordinary managerial skill is the ultimate asset for any organization through which a good strategy is created, roles are fixed and responsibilities are assigned to the right individual and actions are streamlined. This virtue in a manager develops after gaining experience and by good management practices. Because a good manager coordinates with various departments, do brainstorming in order to achieve organizational goals. Suggestions and recommendations from all the individuals (workers and managers) engaged in EMP should also be always welcomed.

### 5.1.5  Implement and Execution of Action Plan

Every organization around the world dealing in energy earns credit for its successful energy program; however, it mostly depends upon the individuals (workforce) of the organization. The role of key authorities at different levels of management (lower, middle and top) is very significant within the organization for successful and effectual execution of the plan on the ground. Sense of responsibility, commitment, capability, integrity, accountability and awareness of all the staff members from top to bottom engaged in the project ensure achievement of goals. Moreover, the technical points of action plan can be implemented by considering the following:

1. There should be proper communication of information about EMP for the targeted audience. Key audience is primarily identified, and then it propagates only the required information, which can be adapted by the audience. Managers, employees, and the stakeholders could be aware of initiative and goals decided for energy performance, only by effective program. Then they all will be well aware of their duties and responsibilities related to project implementation. For common public, energy management program is very important as they are unaware of the fact that their activities and action can affect the energy usage and environment. The new employee of the organization should be trained for spreading awareness about EMP. Informative, creative and attractive posters, pamphlet, leaflets on the table, notice board and in toilets are always considered very effectual tools for awareness. Some employees of the organization managing a facility, have little bit knowledge of energy performance, therefore awareness about facility energy use can be increased by using facts about that facility energy use, operational cost of equipment and impact of energy use on environment. The information must be provided over the tag of the performance equipment that an employee on the job regularly used. There are some official and officers in the organization, who are not directly the part of EMP. Increasing awareness among them will be beneficial for the organization.

2. Support for the EMP initiative is needed at all levels of organization and a good manager is known to be expert in taking such initiatives. He builds capacity through technology, adopting effective practices, procedures, and refresher training.

3. Investment by any organization for training is very essential for successful action plan and to build capacity for the organization as a whole. A trained employee always willing to contribute knowledge, ideas, proper operation of equipment, procedures, give feedback to the managers in upper level of management, that help them to make better decisions in future, and everything which he learned to make successful action plan for EMP. Conducting various training programs vary organization to organization depending upon the nature and type of organization and organizational action plan. However, some commonalities in all training programs are training of procedure and operation, training of administration, and specialized training, etc. In the age of information technology, management information systems play very important role in training programs. It is an advanced mean of sharing information regarding technology, best practices, and operational guidance. The MIS are easily centralized and easily available through an intranet site or a big database.

4. Motivation has been considered as fuel for an organization. Because if personnel are not motivated, they can't work efficiently, which result them to be considered as liability for the organization rather than assets. A good energy manager always

motivates his/her staff by creating some incentives, internal competition, recognition, financial bonus, and prizes for the improvement of energy performance.

5. A system is also developed to regularly track and monitor the progress of EMP. All the activities of energy program are monitored through this system. Everyone can access this system to get information about EMP's established deadline, milestones, and targets. Necessary and essential steps, corrective actions can be taken to improve the overall performance. Energy goals can be achieved by reviewing periodically the information on MIS.

### 5.1.6 PROCESS OF EVALUATING PROGRESS

The evaluation of the data of energy consumed and carrying out activities to meet the demand of performance goals is very much necessary. The established goals are compared with present performance data, then accordingly tasks are accomplished. The energy data is continuously tracked along with the cost of data. A report needs to be prepared for tracked and monitored data and performance, it helps to the authorities to analyze the achievements in energy efficiency with the help of performance matrices incorporated in the report. Energy performance is also evaluated on parameters like degree of improvement in environment, monetary saving and finally the performance must be compared to the other stakeholders in energy sector market for improvement of the rank.

After doing this much of exercise, the management should review the action plan, to understand what work is done up to the mark and where the organization is lagging behind from established goals and should underline the factors that may change results to the set goals. Period evaluation of energy performance helps an energy manager to take cognizance of effectiveness of the implemented program/project, to take better decisions for the upcoming energy projects, to recommend the name of team or individual for award to complete the task on deadline, to document the opportunities through which cost could be reduced for the current and upcoming energy project, document all the best practice for the communication purpose throughout the organization. Those point where goals are not met, corrective measures, preventive actions, and guidelines are decided for further work. The review process can be accomplished by an energy manager by getting feedback from his team, staff members, and from other associated departments engaged in project, by assessing changes in organizational behavior during the energy project implementation, for example, (new renewable energy project implementation create excitement among the employees) by finding out those factors that contributed to missing targets, by quantifying and identifying side benefits like employee's comfort during the project, improvement in productivity, sales patterns, reduction in maintenance and operations cost, and the quality of public relations.

Review of action plan is an effectual exercise and give ideas for new technologies, programs, actions, identify the activities which were felt destructive for EMP can be avoided ensure usefulness of administration and tracking mechanisms, provide information of outstanding financial gain, success stories that build up faith among the stakeholders.

### 5.1.7 ACHIEVEMENT RECOGNITION: A TOOL FOR MOTIVATION

To keep every employee on motivation track is an important and very much an essential responsibility of an organization. The employees put their efforts to pave the way for organization to achieve its goals. They should be recognized time to time. Motivation can be provided monetarily, by awards, gamify them, by trust them, let them know about their importance and purpose of individual, motivate individually rather than team as a whole. This helps in maintaining momentum in energy management program and other staff members also gets motivated from such kind of provision and the energy program gets exposure positively. The validation of the energy management program from any national and international energy body have a great importance for stakeholders who invest directly or through foreign portfolio investment.

## QUESTIONS

Q.1.  State the requirements to have a proper management program of energy.

Q.2.  In order to achieve success in the management of energy, why is the support of top management needed?

Q.3.  Illustrate the process involved in the force field analysis.

Q.4.  Is energy policy needed in an industry? Give reasons.

Q.5.  Explain in brief where the location of an energy manager should be in an organization.

Q.6.  Explain the role of the top management in the management of energy.

Q.7.  Illustrate the process of accountability at different levels in a system of management of energy

Q.8.  Based on your experience, list ways to motivate employees.

Q.9.  Explain the need of proper energy action planning.

Q.10. Explain with reasons why it is essential to undergo training in order to have proper management of energy.

Q.11. Illustrate the advancement of energy information systems. For example, take your industry and provide the list of data needed for monitoring effective energy management.

Q.12. Explain the constraints related to the use of information systems of energy.

Q.13. How much important is goal setting in an energy management organization?

Q.14. What are the steps needed to be followed for formulation of an action plan for energy project?

Q.15. What are the various means of communication of spreading awareness about energy management program?

Q.16. What are the different methods of building capacity in an energy organization?

Q.17. Explain the role of tracking and monitoring system in an energy management organization?

Q.18. How much important is goal setting in an energy management organization?

Q.19. How the progress evaluation process opens new pathways to enhance the energy efficiency?

Q.20. Elaborate the points, an energy manager should keep in his mind during project implementation?

Q.21. Which analysis is useful for avoiding risk in an energy management program?

# 6 Monitoring and Targeting

## 6.1 INTRODUCTION

Monitoring of energy and targeting is an approach to control the use of energy and reduce the wastage of energy for current level of usage by improving the operating procedure. It is based on the principle of measurement and control. Its main motive is to combine the concept of energy usage and statistics.

Generally, monitoring focuses on the existing consumption of energy and using management techniques for reducing such energy consumptions by targeting desirable energy consumption level.

Monitoring is one of the management techniques where essential facilities, such as production of steam, consumption of fuel, energy used in refrigeration, compressor, etc., are monitored in a plant. After monitoring, manager searches the saving opportunities and sets a target to reduce up to that limit. Hence, another technique is targeting, where achievable targets are fixed in a systematic way. Each of the energy cost center utilities is being looked through and observed for comparison with the amount of production and other essential parameters. With the help, such data target and planning can be done to find out the anomaly in the data, actions for remedy, and finally implementing.

Both techniques are widely used and proved to be very effective. By applying these techniques, energy cost can be reduced annually in the range of 5%–20% in various industries/plants.

The ways to increase the performance of energy for a company is illustrated as follows:

- Identifying where energy saving is possible using energy audit or any other survey, walk-through inspection of various plant/industries utilities and the process undergoing in a plant physically;
- Creating awareness and motivating employees for involvement in the management activities related to energy consumption/transformation;
- Practicing techniques related to organizational improvements, such as assigning responsibilities related to energy usage and conducting training.
- Improving the current facilities, where the processes taking place and observation of technological efficiency; and
- Implementing efforts for operation and maintenance for more efficient energy usage.

Monitoring and targeting (M&T) do not include these features and is not a replacement for these actions. It provides information related to the practical approaches and making the system more efficient. It ends with the "accountability loop" by giving a response about the amount of improvement and performance measures taken; up to the level of return from the investment concerning the savings being made. Its usage is rising in Canada and is being used for a long time in the UK since 1980. There are about 50 industries, which have benefited from monitoring and targeting, that includes:

- Saving of energy cost up to 5%–15%, with gradual reductions in GHG emission and other pollutants;
- Implementation of energy management policy, by setting target for initiation of maximum usefulness and savings opportunities for a long time;
- Improvement of the servicing and cost of products by defining the energy product content and service.
- The budget increased due to better insight into energy use in the future for various activities;
- Improvement of maintenance by increasing performance data more available for energy systems;
- Increase in the quality of products, by taking better control of the process related to production; and
- Avoiding waste, by the expansion of the monitoring and targeting principles for the management issues related to environment, such as consumption of water, plant downtime, materials management, and so on.

The monitoring and targeting is a department of management information system for handling various information related to energy supply and usage for gaining energy efficiency.

### 6.1.1 A Case Study: Toray Textiles (EUROPE) Ltd.

Bulwell's Toray Textiles factory (Europe) has been a nineteenth century factory specialized in textiles. The factory manufactures approximately 20 million meters of fabric per annum, with a £40 million sales value, it accommodates about 200 workers. The processes of the factory include bleaching, scouring, and dyeing; three open-width scouring machines, two stenters, and 18 pressurized jet dyeing machines are used.

Toray Textiles procured the factory in 1989 and started a noteworthy energy productivity campaign. During the year 1989–1993, the factory accomplished a decrease in energy consumption of 27%, with half of the investment funds being created through measures which required next to zero capital consumption.

The new program on the management of energy was based on:

- Monitoring and targeting;
- Working employers motivation and training;
- Keeping tracking and focusing; Good maintenance;
- Modern techniques related to management; and
- Investment.

In this new management program, the site manager was in charge of gathering the energy utilization information. In this newly created program, its main objective was calculating the energy consumption for energy efficiency.

Participation of all the working members was found crucial to get success in the program of energy saving. The groups in the company that was dependent on administration and labor union proposed to talk frequently about the issues and provide enhancements regarding energy efficiency. The applied method had shown significant effect on reduction in energy usage, specifically through excellent housekeeping and the recognizable proof of the accompanying real energy sparing undertakings. By December 1993, the energy administration activities executed by the company resulted in the saving of around 75 million kWh. This compares to around one single year's energy utilization for the company and is critical as far as the organization's working expenses and benefits are concerned.

## 6.2   ELEMENTS OF TARGETING & MONITORING SYSTEM

The essential parts of the monitoring and targeting system are:

- **Recording**—Keeping a record and measuring the consumption of energy.
- **Analyzing**—Analyzing the usage of energy to some output of measurement, such as the amount of production.
- **Comparing**—Comparing consumption of energy to some and particular standard or some benchmark.
- **Setting Targets**—Target setting for control in the consumption of energy.
- **Monitoring**—Comparing the consumption of energy to some fixed target for further reduction.
- **Reporting**—Reporting the obtained results for any anomaly related to the target.
- **Controlling**—Applying managerial procedures for taking control of the anomaly occurred.

Particularly M&T system will involve the following:

- **Checking** how much accurate is the invoice of the energy.
- **Allocating** cost of energy department wise.
- **Determining** the level of energy efficiency.
- **Recording** the usage of energy in order to check the improvement of energy consumption.

## 6.3   ANALYSIS OF DATA AND INFORMATION

Bill of fuel and electricity should be collected after every defined period and should be assessed as illustrated. A general method for keeping track of plant is shown in Table 6.1.

**TABLE 6.1**
**Plant-level Information Format**

| Month | Thermal Energy Bill | | | | Electricity Bill | | | | Total Energy Bill |
| | Fuel a | Fuel b | Fuel c | Total in Rs. (Lakhs) | Day (kWh) | Night (kWh) | Maximum Demand | Total in Rs. (Lakhs) | Rs. (Lakhs) |
|---|---|---|---|---|---|---|---|---|---|
| 1 | | | | | | | | | |
| 2 | | | | | | | | | |
| 3 | | | | | | | | | |
| 4 | | | | | | | | | |
| 5 | | | | | | | | | |
| 6 | | | | | | | | | |
| 7 | | | | | | | | | |
| 8 | | | | | | | | | |
| 9 | | | | | | | | | |
| 10 | | | | | | | | | |
| Sub Total | | | | | | | | | |
| % | | | | | | | | | |

## 6.4  APPLICATION

### 6.4.1  ENERGY AND PRODUCTION

Energy used in production process, generally changes the physical or chemical properties of the material. Hence, it is not practically possible to predict and conclude, since the process is of wide variety and complex. However, a theoretical analysis of a few processes results in the relation of energy vs. production into the linear plot:

$$y = mx + c$$

$$\text{or} \tag{6.1}$$

$$\text{Energy} = \text{Slope} \times \text{Production} + \text{Intercept}$$

The intercept is denoted by c, and slope by m is empirical for analysis of the situation and behavior of the system. While there may be some cases where the above given situation is not applicable, then the situation is of a multivariable or a nonlinear function, but the relation above is applicable for majority of the cases.

### 6.4.2  DEVELOPMENT OF ENERGY PERFORMANCE MODEL

The data regarding usage of energy is very useful for learning the energy system, its type, chances of improvement of efficiency, and controlling of the future energy usage. While converting the data into important information, the following steps should be followed:

The initial approach in this study is to find the relationship between the usage of energy and the main depending factors, and relations like Eq. (6.1).

### 6.4.2.1   Step 1: Plot Energy Usage vs. Production

The so-called "scatter plot" of energy usage-production data can be prepared using an excel spreadsheet. It can be seen in Figure 6.1.

Any obtained data can have an error due to the measurements done or the instrument's accuracy during measurements. However, there are other important factors at work, other than "noise." The difficulty lies in distinguishing such factors and the "noise."

The critical question related to monitoring is, why, for the same production quantity, A and B points have quite different consumption of energy? What is to be done to sort it out; the technique used is called cumulative sum control chart (CUSUM) study. CUSUM is a method for finding whether a certain level of energy usage measured changes hugely from the expected value; such changes are random "noise."

The "specific energy" or consumed energy per ton is found. It is a widespread practice, but the effectiveness behind such finding is less.

### 6.4.2.2   Step 2: Determine the Baseline Relationship

The energy utilization and reduction model for every possible value set are:

$$\text{Electricity consumption/Weekly}\left(kWh\right) = 476.48 \times \text{Production}\left(tonnes\right) + 59,611$$

$$(6.2)$$

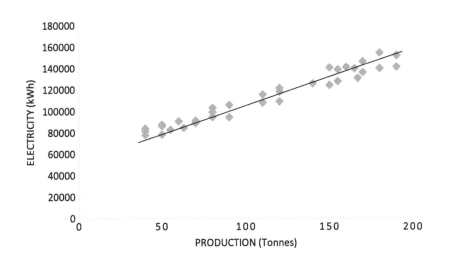

**FIGURE 6.1**   Linear regression of entire data set.

There are two parameters in this model:

- The slope, 476.48, shows the increase in consumed energy per ton of produce;
- The y-axis intercept, 59,611, shows the "no-production" usage of energy, or baseload. It can be seen in Figure 6.1.

The parameters are physically useful for understanding the consumption of energy in the plant and finding the possibilities of increasing the efficiency. It is vital to find the mean energy intensity for such a case; kWh/ton does not correctly denote the performance since it decreases with increase in production and however, the load on which baseline is taken is calculated for many production values.

The value of $R^2$ denotes how fine the points in the trend line fit; 1.0 is a correct fit. $R^2$ of 0.75 or higher values provides a greater level of correctness for the energy performance model created.

Although, the use of this idea is very limited, while taking into account the earlier studies and management control if possible one or more performance change for a period is being analyzed.

What we need is the time duration of consistent performance, which can be used as a baseline or a basis for comparing all other time duration.

Finding a useful baseline generally is a repetitive situation where the basic line regression—CUSUM series is continued repeatedly unless a critical relationship is found. Moreover, being aware of the plant scenario can lead to a place of the specified time duration of consistent performance and the main requirement for a baseline. Which means that no improvements related to the efficiency of energy were created, occurring of any disturbances and the rate of production was usual for the plant, then that time duration can be taken as a good baseline.

At the initial point, the first and third set of the data, which is the oldest part of the data, is sometimes used for the first step at baseline searching. For example, in Table 6.2, assuming that the 1st, 11 weeks, displayed regular and constant output.

$$\text{Consumption of electricity/Week}\left(\text{kWh}\right) = 515.8 \times \text{Production}\left(\text{tonnes}\right) + 60,978$$

$$(6.3)$$

The comparison of two different performance models are shown in Table 6.3.

The information related to incremental load and the baseload for the entire set of values which are less than the values related to the baseline that indicating performance increase occurring after the baseline period. Studies using CUSUM are very much detailed for such a case.

## TABLE 6.2
## CUSUM Data for Production Example

| | Measured Data | | | | Baseline | |
|---|---|---|---|---|---|---|
| Week | Production (T) | Specific Energy (kWh/T) | Total Energy (kWh) | Predicted Energy (kWh) | Difference (kWh) | CUSUM (kWh) |
| 1 | 150 | 938 | 140,726 | 138,020 | 2706 | 2706 |
| 2 | 80 | 1290 | 103,223 | 102,250 | 973 | 3679 |
| 3 | 60 | 1513 | 90,764 | 92,030 | −1266 | 2413 |
| 4 | 50 | 1751 | 87,567 | 86,920 | 647 | 3060 |
| 5 | 170 | 862 | 146,600 | 148,240 | −1640 | 1420 |
| 6 | 180 | 860 | 154,773 | 153,350 | 1423 | 2843 |
| 7 | 120 | 1013 | 121,575 | 122,690 | −1115 | 1728 |
| 8 | 40 | 2036 | 81,436 | 81,810 | −374 | 1354 |
| 9 | 110 | 1051 | 115,586 | 117,580 | −1994 | −640 |
| 10 | 90 | 1177 | 105,909 | 107,360 | −1451 | −2091 |
| 11 | 40 | 2098 | 83,916 | 81,810 | 2106 | 15 |
| 12 | 50 | 1725 | 86,272 | 86,920 | −648 | −633 |
| 13 | 140 | 899 | 125,892 | 132,910 | −7018 | −7651 |
| 14 | 155 | 897 | 138,966 | 140,575 | −1609 | −9260 |
| 15 | 165 | 848 | 139,922 | 145,685 | −5763 | −15,023 |
| 16 | 190 | 801 | 152,274 | 158,460 | −6186 | −21,209 |
| 17 | 40 | 1945 | 77,788 | 81,810 | −4022 | −25,231 |
| 18 | 55 | 1504 | 82,711 | 89,475 | −6764 | −31,995 |
| 19 | 150 | 829 | 124,317 | 138,020 | −13,703 | −45,698 |
| 20 | 80 | 1183 | 94,677 | 102,250 | −7573 | −53,271 |
| 21 | 63 | 1343 | 84,628 | 93,563 | −8935 | −62,206 |
| 22 | 110 | 982 | 108,041 | 117,580 | −9539 | −71,745 |
| 23 | 70 | 1273 | 89,115 | 97,140 | −8025 | −79,770 |
| 24 | 170 | 802 | 136,388 | 148,240 | −11,852 | −91,622 |
| 25 | 190 | 744 | 141,428 | 158,460 | −17,032 | −108,654 |
| 26 | 160 | 883 | 141,215 | 143,130 | −1915 | −110,569 |
| 27 | 120 | 986 | 118,319 | 122,690 | −4371 | −114,940 |
| 28 | 190 | 803 | 152,506 | 158,460 | −5954 | −120,894 |
| 29 | 80 | 1241 | 99,267 | 102,250 | −2983 | −123,877 |
| 30 | 90 | 1050 | 94,468 | 107,360 | −12,892 | −136,769 |
| 31 | 180 | 779 | 140,188 | 153,350 | −13,162 | −149,931 |
| 32 | 70 | 1304 | 91,262 | 97,140 | −5878 | −155,809 |
| 33 | 50 | 1565 | 78,248 | 86,920 | −8672 | −164,481 |
| 34 | 155 | 826 | 128,005 | 140,575 | −12,570 | −177,051 |
| 35 | 167 | 784 | 131,003 | 146,707 | −15,704 | −192,755 |
| 36 | 120 | 910 | 109,192 | 122,690 | −13,498 | −206,253 |

**TABLE 6.3**

**Comparison of Energy Performance Models—Total and Baseline**

| Model | Incremental Load (kWh/ton) | Baseload (kWh) |
|---|---|---|
| Total data set | 476.48 | 59,611 |
| Baseline | 515.8 | 60,978 |

## 6.5 CUMULATIVE SUM CONTROL CHART

CUSUM is a useful method in generating information related to the management of energy for a plant-based on performance and energy consumption system, for example, oven or furnace. Example distinguishing between important events changing the faults related to performance and reduction in noise.

CUSUM stands for "Cumulative SUM of differences," where "difference" is the differences among the original usage and the usage, which is expected due to some pattern that is known as the performance of the energy model. For continuing consumption following the path created, the divergence of the original and created consumption pattern should be less and random in plus or minus value. The net addition of such divergences in time, CUSUM, will be close to 0.

Once there are changes in path occurring because of some fault improvement monitored in the process, difference distribution about zero tends to be less symmetrical and their net addition, CUSUM, varies with time. The CUSUM graph is hence made of linear parts having kinks in between; each kink denotes the change of path; each linear part is time-related during the stable path.

### 6.5.1 STEP 1: CALCULATE THE CUSUM

The relation of the baseline is used for finding the estimated consumption of energy for any level of production. It denotes the difference between this and the original data, which is of principal importance for this study.

Equation (6.1) is for finding the estimated energy usage every week by replacing production every week in the equation. For example, if the production is around to be 150 tons.

$$\text{Energy} = 60,978 + 515.8 \times 150 = 138,348 \text{ kWh}$$

Eliminate the original consumption from the estimated consumption every week to predict and to get the difference. Calculate the net difference for all weeks for the current scenario to get CUSUM.

## 6.5.2   STEP 2: INTERPRET THE CUSUM GRAPH

The critical point on the CUSUM graph is the changes in slope. We see that the changes in slope take place in successive weeks.

Specifically the process being studied and the graph shows:

- There were two steps regarding the reduction of energy; one in week 12 and one in week 18.
- The first one did a savings of 73,500 kWh and the other 36,800 kWh by the time of week 25, the breaking down of the second one is done.
- This second one was back on week 30 and followed by the end; the measurement resulted in savings of 201,300 kWh.

## 6.5.3   SUMMARY: REGRESSION AND CUSUM

Evaluation:

- Evaluation of energy use vs. Variable of independent nature, which is applicable.
- Do the regression analysis for determining a relation in terms of function between energy usage and the variable of nondependent nature.
- Finding the baseline for the relation related to regression.
- By using the baseline to estimate "predicted" usage of energy for actual values of the nondependent variable.
- Find the error between estimated and the actual consumption of energy, and the cumulative sum of differences (CUSUM).
- Draw and analyze the CUSUM plot.

All of these are possible only with the basic knowledge of the Excel.

## QUESTIONS

Q.1. Explain the difference between monitoring and targeting.
Q.2. Describe in brief the main elements of monitoring and targeting system.
Q.3. Discuss in brief, the benefits of monitoring and targeting systems.
Q.4. Explain the term "benchmarking" and its benefits.
Q.5. Discuss the internal and external benchmarking differences.
Q.6. Explain the different varieties of energy studying and focusing systems based on the industry.
Q.7. Illustrate the primary 10 processes of "Monitoring and targeting" that needs to be undertaken as an energy manager.
Q.8. Illustrate how a CUSUM chart is drawn with an example.
Q.9. Why is the CUSUM technique most useful?
Q.10. Explain the ways to handle energy consumption and production in a plant.

# 7 Electrical Energy Management

In electricity billing, all the industries pay based on two-part tariff in which customer pay for maximum demand (kVA) and electricity consumed (kWh) during the period.

## 7.1 MAXIMUM DEMAND BASICS

Bill for maximum demand is nothing but the fixed charge of capacity blocked provided for consumer's needs. A tri-vector meter installed at the consumer end records. The maximum demand used by a consumer in billing duration is recorded by a tri-vector meter installed at consumer's firm, apart from another important consumption component like active power (kWh), reactive power (kVArh), apparent power (kVAh), and power factor (PF). At present, in many places, a "time of day tariff" process is used instead and charged consumer variable rates for maximum demand drawn during different times of the day. It is seen, there could be lower rates for night hours when the lightly loaded utility is used, and higher rates are noted during evening hours as utilities are stretched to meet maximum load demands. With the help of this study of the load curve patterns is done and based on prevalent tariff structure provisions, optimize maximum demand to save on maximum demand charges.

Load curve of a plant: The demands vary with time due to the change in user consumption and the usage of electricity. The measurement is done based on a particular time, and the average values calculated are being represented by the dotted line in the horizontal plane.

Load management strategies: The technique used for the optimization of the maximum demand is given below:
- Loads reschedule;
- Motor load staggering;
- Products storage in the various processes of materials such as refrigerating;
- The non-important loads are being shaded;
- Using a captive power plant, and
- Compensation of reactive power.

Rescheduling and staggering loads: The left-hand displays the consumer demands before the shifting of load and the right hand represents the after effect of the shifting of load.

Products Storage in material processing utilities: There are some plants where studies show that they have extra capacities are present for a few products and process. Utilities such as pumps, refrigeration, and others need electrical power to run. Such capacities can be used during the period of less demand, and during the peak, time usage can be managed.

## 7.2   IMPROVEMENT OF POWER FACTOR

In recent electrical distribution system, the primary loads are resistive and inductive.

- Resistive loads are generally resistant toward heating and lights.
- The inductive loads are generally motor, transformers, and furnace.

The power requirements of the inductive loads are the active power for doing work and the reactive power for generating and maintaining the electromagnetic field.

The total power is the vector sum of reactive and active power. Power supplied works based on active and reactive power in electrical systems.

The main component of usage is the active power, which is in kilowatts. The advantage of using a power factor being close to unity is the reactive power being close to zero. Hence, from this process, the capacity of the electrical system can be used in an optimized manner having at the lowest loss. The devices sometimes affect the users for low power and help users with high power factors in terms of their bills generated or tariffs. The installation of capacitors helps in the maintenance of the reactive power generated due to the inductive loads; it benefits the user by improving the power factor, which results in proper voltage, decreasing the demand, and reducing the loss due to distribution.

## 7.3   CAPACITOR LOCATIONS

The motors having more the 50 hp should have capacitors with them for power factor correction at the terminal of the motors to reduce the load in the distributor circuit. The first method of compromising will reduce the cost of installation. The cheapest method is the connection of the capacitors at the entrance of the service, and current high feed flows entrance and service endpoint equipment is a useful device.

Increasing the PF can reduce the load due to distribution and the maximum demand within the devices.

Then a flow of current through the conductors results in the heating due to the effect of resistance, known as the distribution loss. In industrial cases, the range of this loss is from 1% to 6%. It is directly proportional to the current squared, which gets reduced with the decrease in the power factor.

$$\text{Decrease of loss in percent} = 100 \times 1 - \frac{\left(\text{Original P.F.}\right)^2}{\left(\text{Desired P.F.}\right)^2}$$

The apparent power when multiplied by the power factor gives us active power. However, to decrease the demand, the power factor is to be decreased around unity. For payback analysis, the reduced max demand charges are found out in terms of the capacitor investment. The analysis of the various data such as monthly bill can help to find out the cost benefit.

## 7.4   PUMP

"A mechanical device, using suction or pressure, to rise or move liquid by mechanical action" is called a pump.

Pumps are available in various sizes, each having numerous applications. The pumps are divided based on the working principle, such as displacement pumps or dynamic pumps. The pumps of dynamics type are divided into special effect type pump and the pump of centrifugal type. The classification of the displacement pump is based on pumps of rotary or reciprocating nature. Based on the working principle, any pump can tackle any liquid by any of the designs. Based on the types of pumps, the centrifugal is the most economical, after which comes the rotary and the centrifugal. The most efficient is the pump of positive displacement type, which has efficiency more than the centrifugal pumps, the advantages of having a high-efficiency balance the high cost related to the maintenance.

### 7.4.1   CLASSIFICATION OF PUMPS

The two main types of pumps are as follows:

1. Positive displacement pumps
2. Rotodynamic pumps

#### 7.4.1.1   Positive Displacement Pumps

It is either reciprocating type or of rotating type.

1. **Reciprocating types**
   The main designs are:
   a.  Piston or plunger type
   b.  Membrane type pumps
2. **Rotary positive displacement pumps**
   The main designs are:
   a.  The pump of gear wheel type
   b.  Worm pumps
   c.  Partition or vane pumps
   d.  Mono pumps

The main specialty of the positive displacement pumps is that it displaces the fixed amount of fluid of given volume with the increment in the size of the compartment of the fluid for the fluid intake and resulting in the reduction in the area of the compartment of the leaving fluid, the positive displacement occurs. The positive displacement pump can create high pressure by only taking control of the safeties associated with the pump at the discharge or delivery. However, the pumps have shallow discharge since the volume swept us very limited based on the displacement and the speed.

Positive displacement pumps are more intricate mechanically, and therefore, tend to be more expensive. It is the sole reason why they are taken as the second selection after the rotodynamic pumps for job completion.

### 7.4.1.2 Rotodynamic Pumps

In this type of pump, the energy is applied at the impeller of the pump in terms of rotary motion. The kinetic energy developed in the wheel then transfers to the pumping fluid via the blade known as the fluid velocity. Hence, the kinetic energy developed is converted to pressure energy by the non-moving parts of the pump.

There are three types of rotodynamic pumps:

1. Axial flow;
2. Mixed radial and axial flow; and
3. Centrifugal or radial flow (Figure 7.1).

The flow patterns are as follows:

1. Pumps of centrifugal type of radial flow type where the fluid goes out from the impeller in a direction radially.
2. The propeller air axial pump flow where the flow of the resultant fluid is in the axial direction.
3. The mixed flow where the leaving fluid leaves in the combination of axial and radial flow.

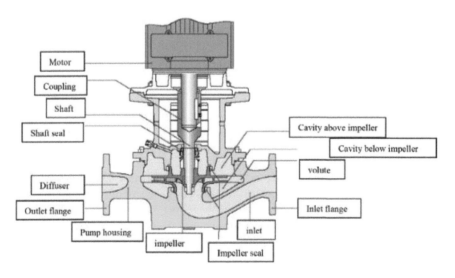

**FIGURE 7.1**  Centrifugal pump. (From Raj, P.P. et al., *Mater. Today*, 21, 175–183, 2019.)

### 7.4.1.3 Other Types of Pumps

The other types of pumps are as follows:

1. **Positive displacement types**

   The main characteristics of these types of pumps are its low discharge but having very high pressure. The pressure releasing valves is usually available on the side of the delivery side in for safety measures due to the high pressure and to limit the maximum pressure. When these pumps are made on the concept of multistage or centrifugal or mixed type of flow, they can discharge at high pressure, which would or else need a pump of positive displacement type (Figure 7.2).

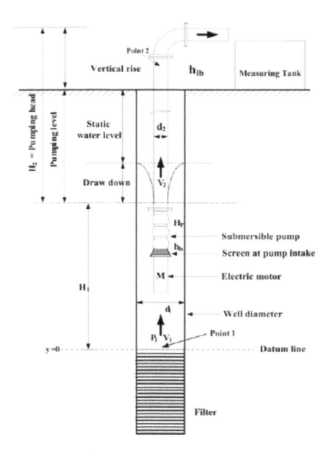

**FIGURE 7.2** Submersible pumps. (From Haque, M.E. et al., *Energy Procedia*, 160, 123–130, 2019.)

2. **Pump drawing water from a free surface**

Only a low lift is possible. They are:

a.   Water wheels

b.   Archimedean screws

**Finding the efficiency of the pump**:

1.   $(P)$ Consumption of energy $(\text{kw}) = \sqrt{3VI \, \text{Cos}\Phi}$

2.   Efficiency of the combined motor plus pump

$$= \frac{M \times 9.81 \times TDH(mWC)}{P} \times 100$$

where:

P = Power input (kW)

m = Flow rate (m³/sec)

TDH = Net head (differential) (mWC)

$$\text{Pump efficiency } (\%) = \frac{\text{Combined eff motor}}{\text{Motor eff}} \times 100$$

3.   Flow rate $\left(m^3/h\right) = \dfrac{(P)(\eta_p)(\eta_m)(3600)}{(TDH)(9.81)}$

where:

$\eta_p$ = Pump efficiency

$\eta_m$ = Motor efficiency

4.   Power consumed $(\text{specific}) = \dfrac{\text{Power consumed } (\text{kWh})}{\left(\dfrac{m^3}{h}\right) \text{Flow rate}}$

3. For detailed analysis, shutting the head for trial can be done for maximum a minute interval just for ensuring the full closure of the discharge valve without passing. If the given head is 90%–95% less than the operating conditions, some problems arise such as:

   • A leak of the gland seal,
   • Misalignment of the axis of the shaft,
   • Pitting or wearing out of the impeller,
   • Wearing out of the case, and
   • Wearing of the bearings.

4. Based on the current efficiency of the pump, find the reduction in cost for:

   • Replacing the pump with a pump of higher efficiency or
   • Replacing the impeller.

5. Based on the percentage of profits of the various parameters, such as motor flow, head or input, the benefit of cost is calculated for the application of:

   • Variation in the speed drive,
   • Optimizing the impeller size or
   • Small applications related to multi-pump.

## 7.5 FAN SYSTEMS

Fans supply air to ventilate and for other processes required by the industry. Fans operate by generating pressure and creating the movement of air as opposed to the resistance offered by the ducts and dampers. The fan receives its energy from the transmitting shaft of the rotor, which supplies it to the air.

### The difference in fan compressor and blowers

The man difference associated with both devices is how the movement of air happens and by the system pressure they must operate against. ASME rules state that the discharge ratio to the suction pressure is used for the evaluation of the fans and blower. The rise in pressure for a fan is around 1136 mmWg. For blowers, it is from 1136 to 2066 mmWg, and in the case of compressors, it is more than the blowers.

### 7.5.1 TYPES OF FAN

The fans are selected based on the flow volume and pressure. The efficiency of the fans varies based on the design. The main two categories are the centrifugal types and the axial types. In centrifugal type, there is a change in the flow of air two times during the time of entering and leaving. However, for the axial, the fans leave with no change in the direction of air (Figures 7.3 and 7.4).

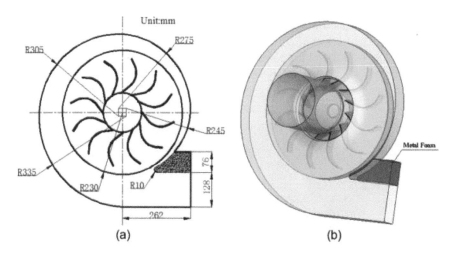

**FIGURE 7.3** Centrifugal fan. (a) Two-dimensional side view and (b) three-dimensional view. (From Xu, C. and Mao, Y., *Appl. Acoust.*, 104, 182–192, 2016.)

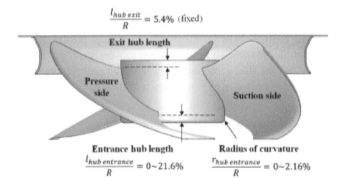

$$\frac{l_{hub\ exit}}{R} = 5.4\% \text{ (fixed)}$$

**Exit hub length**

**Pressure side**

**Suction side**

**Entrance hub length**

$$\frac{l_{hub\ entrance}}{R} = 0\sim21.6\%$$

**Radius of curvature**

$$\frac{r_{hub\ entrance}}{R} = 0\sim2.16\%$$

**FIGURE 7.4**   Axial fan. (From Jung, J.H. and Joo, W.-G. et al., *Int. J. Refrig.*, 101, 90–97, 2019.)

### 7.5.1.1   Types of Centrifugal Fan

The main varieties of centrifugal types are as follows:

1. Radial forward-curved
2. Backward curved

The fans of radial types are mostly used in the industry because of its very high pressure in the static state or the static pressure. They can manipulate or change in the broad stream of air. The designs of such types of fans are straightforward; hence, they can be utilized at a very high temperature with the blade speed being medium.

The forward-curved fans are used for the cleaning of the environment and have a low operating temperature. These types of fans have low speed at the tip, and flow of air through the fan is very high. Hence, generally, they are used for the movement of large volumes of air at a slight relative pressure.

However, the backward curved fans are much more efficient than the forward-curved fans. The backward curved fans work on the peak level of its power consumption, and its demand for power is well-utilized in the range of the flow of air. The fans of backward curved flow are said to be "non-overloading," since which the change in the pressure at the static condition does not result in overload (Figure 7.5).

**FIGURE 7.5**   Forward curved and backward curved. (From Bai, Y. et al., *Int. J. Hydrogen Energy*, 42, 18709–18717, 2017.)

### 7.5.1.2   Types of Axial Flow Fans

The main types of fans of axial flow are as follows:

1. Tube-axial,
2. Vane-axial, and
3. Propeller.

The fans of the axial tube have a wheel inside them, which is enclosed by the housing of cylindrical type, and the clearance is very less among the blade and the housing to increase the efficiency of the airflow. The wheel runs faster than the fans of propeller type resulting in high operation of around 250–400 mmWC with the maximum efficiency around 65%.

The fans of axial van type are more or less like axial tube types but having extra guide vane for the improvement in the efficiency by making the airflow straight. Hence, it has higher static pressure and does not depend much on the static pressure of the ducts. These fans are generally used for pressure up to the level of 500 mmWC. The axial vane fans are the most energy-efficient fans, and they are to be used wherever possible.

The fans of propeller type work at less speed and the working temperature is moderate. There is a massive change in the flow of air with very fewer changes in the pressure at the static condition. This type of fan handles a large volume of air at low pressure, or the delivery is free. These are used in the indoors as exhaust fans. For the fans used in the outdoors, its applications are an air-cooled condenser and cooling tower. Its efficiency is very less, which is around 50% or less.

### 7.5.2   Assessment of Performance of Fans

The testing of fans is generally done based on the flow of air, head or the draft, and fan temperature and motor of electrical type in kW input on the motor side.

#### 7.5.2.1   Measurement of the Flow of Air

1. **Static pressure**: Fan static pressure is the resistance pressure the fan has to blow against in order to move air in the desired direction. If the fan is blowing against a high resistance pressure, it requires more horsepower and delivers less air.
2. **Velocity pressure**: It is the pressure in the direction of flow due to the air flowing through the duct. It is used for the calculation of the velocity of air.
3. **Net pressure**: Net pressure is the addition of static pressure and the velocity pressure. Both pressure can change pressure according to the change in duct size. However, the total pressure remains the same and it varies with friction. The flow of a fan can be measured by using anemometer, air flow sensors, and combination of Pitot tube and manometer.

**Measurement by Pitot tube**:

The inner part measures the total pressure, and the outer tube measures the static pressure. On connecting the inner and the outer end of the tube of the manometer, the velocity pressure can be obtained (Figure 7.6).

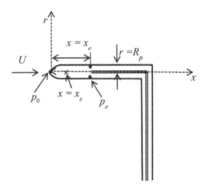

**FIGURE 7.6**   Pilot tube measurements. (From Chevula, S. et al., *Flow Meas. Instrum.*, 46, 179–188, 2015.)

### 7.5.2.2  Velocity Pressure/Velocity Calculation

During the calculation of the pressure velocity, the diameter of the duct is taken into account and is to be measured correctly. It helps in the calculation of the velocity and air volume in the duct. In generally all of the cases, few measurements are to be taken for the system at a particular point.

- With change in the duct size, pressure velocity changes. Due to friction, air velocity is found higher in the center whereas the velocity is found lowest near the wall. By changing the duct configuration like bent and curve, the velocity decreases. The accurate place for measurement in a section of duct that is straight for at least 3–5 diameters after any elbows and branch entries.
- For finding the average velocity, 4–5 readings of velocity pressure is to be done. For each of the cases, the velocity is to be found out, and the average velocity reading is to be used.

1. **Air density calculation**: Initially, the density of air is found out. The air density is required to find out the velocity and velocity volume from velocity pressure measurement and it is dependent on height and temperature:

$$\text{The density of the gas } (\gamma) = \frac{273 \times 1.293}{273 + t^\circ C}$$

where $t^\circ C$ is the gas temperature or temperature at the site.

2. **Calculation of Velocity**: On establishing the density of air and velocity pressure, the velocity is obtained from the following equation:

$$\text{Velocity v, (m/sec)} = \frac{cp \times \sqrt{2 \times 9.81 \times \Delta P \times \gamma}}{\gamma}$$

where:

cp = constant of Pitot tube, 0.85 provided by the manufacturer,

P = Avg differential pressure found by the Pitot tube calculated at various points through the entire cross section, and

γ = Air or gas density during the testing condition.

3. **Calculation of volume**: The duct volume is fond out from the equation:

$$\text{Flow rate of the volume } (Q), \text{m}^3/\text{sec} = \text{Velocity} \left(\text{m/sec}\right) \times \text{Area} \left(\text{m}^2\right)$$

### 7.5.2.3 Efficiency of Fan

There are two ways to find out the efficiency: mechanical efficiency or net efficiency and static efficiency. Both calculate the conversion of horsepower into pressure and flow.

The equation for finding out the mechanical efficiency of the fan is

$$\eta_{mech} (\%) = \frac{\text{Volume in m}^3/\text{sec} * \Delta p \left(\text{total pressure}\right) \text{ in mmWC}}{102 * \text{Power input to fan shaft in kW}} \times 100$$

The outlet pressure velocity is not included in the static pressure but the rest same: Fan static efficiency,

$$\eta_{Static} (\%) = \frac{\text{Volume in m}^3/\text{sec} * \Delta p \left(\text{static pressure}\right) \text{ in mmWC}}{102 * \text{Power input to fan shaft in kW}} \times 100$$

Driving motor in kW is found out using the analyzer for a load; the product of power in kW and efficiency of a motor is the power of the shaft for the fan.

### 7.5.3 Energy Conservation Opportunities in Fans

The system for developing an efficient energy system involves developing a curve for the system with accuracy, proper selection of points of operation for the fan, and correlating it with the position where the best efficiency is possible. The highest efficiency of selection is done for the fan and achieving an efficient controlling capacity of the fans and correct usage of O&M.

1. **Flow control strategies and energy conservation**: Mostly, when a fan used in the system after design and installation, it works at a particular speed. Only in a few cases, its speed needs to be changed, such as

during the usage of the new duct. The situation where the fan is more prominent in size; then, time is needed for reducing flow air. There are many methods to reduce the flow, such as changing the pulley or damper control.

2. **Damper controls**: Some of the fans have damper controls installed at the outlet or inlet. It acts as a device that can change the volume of air by the addition or removal of resistance. The resistive force allows the fans to move up or down following a particular curve creating air at low or at high speed. The adjustment made by the fans is limited, and not very efficient in terms of energy.

3. **Pulley change**: The change in the pulley of the fans is used when the change of flow volume of air is needed permanently, and the current working fan will be able to handle that change. The change in volume is generally due to the change in speed. The best way of changing the speed is changing the pulley, where a V-belt system is used for the fan. The speed change is done based on drive changing, usually used pulley de-rated having a massive size fan than is required, and the working of the damper keeps in observance at all times.

   a. **Inlet guide vanes**: The guide vanes of inlet type is also used to control the variable airflow; curvature done at areas not directly related to inlet fan opening. In closing, it is in contact with the stream of air. There is an initial swirling of the air before entering the housing of the fan. When there is a change in the angle of the fan blade 0, there is a change in the curve of the fan. The efficiency of the guide vanes is from 100% to 80% reduction in the flow in the modest cases. After 80%, there is a drastic drop in the efficiency of the energy for the fan. The axial flow fans have variable pitch blades that are pneumatic or hydraulically controlled during the stable condition.

   b. **Variable speed drives**: These are very expensive and provide considerable variation in the speed. It is used to reduce the speed of the fan and changing the flow rate of air. The speed of the fan is predicted at a variable speed based on the laws of the fans. However, these variations in the speed of the fans are not useful if the frequency of variation is not proper. The control system efficiency should be considered for the variable speed drive for analyzing the amount of power consumed by the system. The energy efficiency for the flow is done by selecting the proper method of flow control.

## 7.6   ENERGY-EFFICIENT MOTORS

The function of the motors is to convert electrical energy into mechanical energy using the interaction of the magnetic field between the stator and the rotor. There are varieties of electrical motors available in the market such as induction motor, synchronous motor, and DC motor. Every motor has the same four types of components, which are frame, stator, rotor, and bearings. Most of the motors are squirrel cage motors (Figure 7.7).

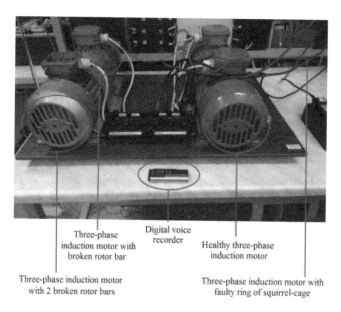

Three-phase induction motor with broken rotor bar

Digital voice recorder

Healthy three-phase induction motor

Three-phase induction motor with 2 broken rotor bars

Three-phase induction motor with faulty ring of squirrel-cage

**FIGURE 7.7** Induction motor. (From Glowacz, A., *Appl. Acoust.*, 137, 82–89, 2018.)

### 7.6.1 SALIENT FEATURES OF MOTOR PERFORMANCE

**Motor speed**: The motor speed is the number of rotations made by the motors in a particular period, generally given by RPM (revolutions per minute). The RPM represents the synchronous speed with the help of an equation given below, and the unit of frequency is given by hertz or cycles/second.

$$\text{Synchronous speed}\,(\text{RPM}) = \frac{120 \times \text{Frequency}}{\text{Number of poles}}$$

It is known that the AC motor speed can enormously vary with the change in frequency. The motor operates at a speed less than the synchronous speed. The synchronous speed and full load speed difference are known as the slip percentage. It is calculated from the following equation:

$$\text{Slip}\,(\%) = \frac{\text{Synchronous speed} - \text{Full load speed} \times 100}{\text{Synchronous speed}}$$

**The relation between voltage and frequency**: The frequency of an inductor is directly proportional to its impedance. When the frequency is low, the impedance is around zero, the circuit behaves as a short circuit. The motor voltage is to be changed to keep the flux constant; it is generally constant over the entire speed range. By constant ratio, the speed, which is fixed can be made to run a speed and it varies by supplying a constant torque. When the speed is less, there is resistance in the windings; hence, the ratio should be changed sufficiently for providing proper magnetic flux for running the motor, where it can be changed using VFD by altering the boost voltage.

**Load versus power factor**: On decreasing the load, the active current gets reduced, although the magnetic current does not change. Hence, there is a decrease in the power factor of the motor as the load applied gets reduced.

**The efficiency of motor vs. load**: The electrical motors convert electrical energy based on the principle that the force acting on the electric current and the magnetic field is being excited by electrical as well. Based on their principle of functioning, it has relatively high efficiency. The efficiency range is around 90%–94% for the standard motors (up to 100 kW). The efficiency of the large motors is higher than the efficiency of the small motors.

$$\text{Efficiency} = \frac{\text{Output}}{\text{Input}} \times 100$$

$$\text{Efficiency} = \frac{\text{Input} - \text{Losses}}{\text{Input}} \times 100$$

$$\text{Efficiency} = \frac{746 \times \text{HP output}}{\text{Watts input}} \times 100$$

It is to be kept in mind that the efficiency is at its peak at 75% load and reduces very much when the load is below 30%.

The energy efficient motors (EEM) are that in which improvements in design are done for increasing the operating efficiency based on the previous standard designs. The improvements in the designs are mainly focused on reducing the intrinsic losses of the motors. The improvements made in the motors are reducing the loss due to silicon steel, and a larger core to increase the physical activity, thick wires to reduce the resistance offered, etc.

In India, the EEM available is 3%–4% higher in terms of performance compared to the standard motors. Following the standards of the BIS, the EEM is generally made to work without a reduction in the efficiency of the motor when the capacity of rating from 75% to 100% creates a significant benefit related to the application of the load. For the power factors, it is nearly the same or higher than the standard motors. In addition, the EEM operates at a temperature less than the standard motors and less noise and high acceleration and very few changes due to voltage fluctuation.

Methods applied for increasing efficiency are:

## 7.6.2 Stator and Rotor I²R Losses

These are the maximum loss that occurred at the inductor, which takes 55%–60% of the net loss. The $I^2R$ loss is generally a heating loss that arises when current flows through the rotor and stator and it is the function square of current and resistance.

Now the conductor resistance is the function of the thickness of conductor material, its cross section, and length. A proper size of the copper material can be used to reduce the resistance. When the current flowing through the motor will be reduced, there will be a reduction in the magnetizing effect due to the current. It results in a reduction in the flux of the current and reducing the gap of air. The $I^2R$ rotor loss is due to the conductor rotor functioning and the slip of the rotor. On using conductors of copper materials, there will be a reduction in resistance due to winding. The synchronous motor speed reduces $I^2R$ motor loss.

### 7.6.3   Core Losses

The loss associated with the core is due to the effect of hysteresis and eddy current due to the 50 Hz magnetic effect of the core occurring in the region of the stator rotor magnetic area in the steel. The losses are not dependent on the load applied, and it amounts to 20%–25% of net loss.

The loss occurring due to hysteresis is generally due to the density of flux. It can be reduced by using laminated silicon steel of low-grade. The density of flux being reduced is achieved by using a large core for the stator and the rotor. The loss due to eddy current is generally because of the circulation, and it can be reduced by using laminations of thin materials.

### 7.6.4   Friction and Windage Losses

The loss due to friction and windage is generally due to friction in the bearings and air circulation through the motor and can have a net loss of 8%–12%. This loss does not depend on the amount of load. The lessening of the generated heat from the stator and rotor allows the usage of a small fan. The loss due to windage also decreases with the diameter of the fan.

### 7.6.5   Stray Load-Losses

This type of loss is directly proportional to the load current square and is formed by flux leakage due to the induction of the current supplied to the load. The laminations and the loss associated is about 4%–5% of the net loss. The processes of reducing the losses are by proper selection of the slot numbers and the gap of air. There is a variety in the rating of these types of energy-efficient motors and their efficiencies at full load are higher up to 3%–7%.

Due to the modifications to increase the performance of the motors, the cost of these energy-efficient motors is very higher than the general motors available in the market. The high cost gives a return value related to the cost of operation and savings and other areas and afterlife replacement. Although if the afterlife replacement is not proper, the cost related to the motors may not be beneficial (Table 7.1).

## TABLE 7.1
## Energy Efficient Motors

**Energy Efficient Motors**

| Area of Power Loss | Improving Efficiency |
| --- | --- |
| 1. Iron | A thin gauge is to be used to reduce eddy current loss. The core of more extended size increases the content of steel hence decreasing the losses due to operating flue density. |
| 2. Stator $I^2R$ | Using more amount of copper and increasing the area of a cross section to increase the windings of the stator will result in a decrease in the winding resistance and loss of flow of current. |
| 3. Rotor $I^2R$ | The large conductor rotor size decreases resistance and increases the current flow rotor. |
| 4. Friction & Windage | A fan that has less loss should be used to reduce the loss due to the movement of air. |
| 5. Stray Load Loss | Using an efficient design and high-quality maintenance process to reduce the loss of energy due to straying. |

### 7.6.6 TECHNICALITY RELATED TO THE MOTORS FOR ENERGY EFFICIENCY

1. The energy-efficient motors have a long durability, and their maintenance is low. The re-greasing time increases due to low temperature. It is observed that for every 10 degrees decrease in temperature, the life of the motor increases twice.
2. An EEM with service factor 1.15 and its design for operation 85% of rated power is good for selection.
3. Problems in the electrical circuit significantly affect the motors and decreases the quality of the motors in terms of operation.

## QUESTIONS

Q.1. Explain power factor usefulness of improving power factor.
Q.2. How can energy conservation be achieved by improving the power factor of an induction motor?
Q.3. A three-phase induction 75 kW motor operates at 55 kW. The measured voltage is 415 V and current is 80 A. Calculate the power factor of the motor.
Q.4. What are constructional (rotor and stator only) effective operational differences between conventional motor and energy-efficient motor?
Q.5. Explain the types of losses in induction motors and their total percentage of loss out of the net losses.
Q.6. Describe the energy efficiency affecting factors for electric motors.

Q.7.   Explain what factor leads to core losses during rewinding.

Q.8.   Illustrate the process by which speed control of the motor is obtained.

Q.9.   State the ways to increase the efficiency of energy for motors.

Q.10.  Explain the loss due to the efficiency of a rewound motor.

Q.11.  A 50-kW induction motor having efficiency when fully loaded of 86% is replaced by a motor of efficiency 89%. How much energy is saved for 6000 h/yr and the energy cost is Rs. 4.50/kWh?

Q.12.  State the difference among fans, blowers, and compressors.

Q.13.  Describe the terms static head and friction head.

Q.14.  Explain in brief the diagram loss of energy during throttling for a pump (centrifugal).

Q.15.  State using a sketch the concept of characteristics of heat flow for a pump and system resistance.

Q.16.  Explain the effects of the pump being over-sized.

Q.17.  Illustrate the energy conservation of energy changes in the pumping system.

Q.18.  Using a sketch describe the types of cooling towers.

Q.19.  Explain "Cycles of Concentration" and its relation to cooling tower blowdown.

Q.20.  Illustrate what can affect the cooling tower performance.

Q.21.  Illustrate the conservation of energy possibilities in a cooling tower system.

# 8 Thermal Energy Management

## 8.1 BOILER

A boiler is a closed pressure vessel that supplies heat energy in order to increase water's temperature to the point that it progresses to form steam. The pressurized steam is then used for giving heat energy in several industrial processes. The volume of steam is increased 1600 times approximately when water gets boiled which delivers power practically explosive as close as gunpowder. The high pressure inside the boiler results in an equipment which requires enormous care.

A system of boiler is made of three parts:

1. Feedwater system,
2. Steam system, and
3. Fuel system.

The feedwater system supplies the boiler with water and controls it consequently to take care of the demand of the steam. Different valves at different places are given for repair and maintenance. The steam system gathers and helps in controlling the steam formed in the boiler. Steam is guided through a channeling framework to the area of utilization. All through the system, the pressure of steam is controlled using valves and checked with pressure gauges at several points. The fuel system incorporates all the necessary materials needed to give fuel to produce the required heat. The machineries needed in the fuel system relies upon the kind of fuel utilized by the system.

The measurement of heating surface of the boiler is generally calculated in square meters. Any component of the boiler material that influences in steam generation is the heating surface. The bigger the heating surface of the boiler, the higher will be its ability for steam generation. There are a few types of heating surfaces and they are as follows:

1. Radiant heating surfaces which includes surfaces that are all backed with water which is directly available to the combustion flame supplying radiant heat.
2. Convection heating surfaces which includes surfaces that are all backed with surfaces prone to gases are not only hot but combustible in nature too.
3. Extended heating surfaces which is found in few boilers of water tube type containing economizers and superheaters.

## 8.1.1   Boiler Types and Classification

Mainly, the industrial boilers found in A SME are of four types:

1. **Fire-tube boilers**: Also known as "fire in tube" boilers have tubes of very long size made of steel where hot gases travel from a heater to the water to be transformed into steam in a circulating manner. It is generally utilized for small steam limits (up to 12,000 kg/h and 17.5 kg/cm$^2$). The merits of fire tube boilers help in reducing the cost of capital and the effectiveness of fuel (more than 80%). But they are difficult to work with, as it acknowledge huge variation of load, and in light of the fact that they can deal with vast volumes of water, and create less variety in steam weight.

2. **Water-tube boiler**: Also called "water in tube" boilers; hot gases flows outside whereas water flows inside the tubes. These boiler drums can be single or multiple which provide high capacities in storage. It can be used to deal with high-pressure steam and have efficiency more than that of the fire tube types. Used in the control plants having variation of steam extending from 4.5 to 120 t/h, and associated with high capital cost. These are used to provide high pressure and high quality of steam with stringent water quality standards.

3. **Packaged boilers**: These types of boilers are constructed as a complete package. When it is installed at site, it requires only piping for the supply of water, fuel supply, and electrical connections to make it operational. Packaged boilers are generally of shell type with fire tube design in order to accomplish high rates of heat exchange by process of convection and radiation. These are classified on the basis of the number of passes. The chamber where the burning occurs is known as the primary, followed by 1, 2 or 3 arrangements of fire tubes. The widely recognized boiler of this type is a 3-pass unit with 2 sets of fire tubes and with the fume gasses leaving through the rear of boiler.

4. **Fluidized bed combustion (FBC) boilers**: In FBC, the ignition of fuel happens on a floating (fluidized) bed. At the point when a uniformly distributed air or gas moves upward through a finely partitioned bed of strong particles, for example, sand upheld on a fine mesh, the particles are not distributed at low speed. As air velocity is steadily increased, bubbles are formed and the particles are floating noticeable all around the stream. Further increments in velocity offer speed to the development of bubbles, high turbulence, and quick blending, and the fluidization of the bed has occurred. A bed boiler of fluidized type provides favorable circumstances of low discharge, high effectiveness, and versatility for utilization of less calorific-value fuels like biomass, municipal solid waste, etc.

## 8.1.2   Boiler Performance Evaluation

The effectiveness of a boiler, which incorporates thermal efficiency and evaporation ratio, degrades after sometime due to poor combustion, fouling of heat transfer region, and deficiencies in maintenance and operation. Poor quality of fuel and water can bring about boiler execution very down even for a new boiler. The boiler effectiveness test has its advantage to compute the amount of productivity from the point of view of design value and identify areas for improvement.

1. **Thermal efficiency**: Thermal efficiency of boiler is characterized to be the ratio of heat input that is effectively utilized to generate steam. Mainly, two strategies are present for surveying boiler effectiveness: direct and indirect. The direct strategy, the proportion of heat yield to supply of heat is found out. In the indirect strategy, all the indirect processes of loss of heat for a boiler are calculated and its productivity is figured out by subtracting the loss from 100. The different losses are computed as demonstrated in the example given below.
2. **Evaporation ratio**: Evaporation ratio (steam to fuel ratio) is one of the conventional and important parameter to observe the daily performance of boilers. For lower capacity boilers, direct methods can be attempted, yet it is desirable over efficiency calculation using indirect method since an indirect process allows assessing of all losses and can be an instrument for decrementation of losses. For the process of direct method, the quality of steam estimation displays vulnerabilities. Particular method is applied for computing and methods for calculation.

### Example: Calculation of Direct Efficiency

Find the direct efficiency of the boiler from the data given below:

Boiler types: coal-fired
Amount of generated steam (dry): 8 TPH
Pressure of steam (gauge)/temp: 10 kg/cm² (g)/180°C
Amount of consumption of coal: 1.8 TPH
Temperature of feed water: 85°C
GCV of coal: 3200 kcal/kg
Steam enthalpy at 10 kg/cm² (g) pressure: 665 kcal/kg, (saturated)
Inlet fed water enthalpy: 85 kcal/kg

$$\text{Efficiency of boiler } (\eta) = \frac{8 \text{ TPH} \times 1000 \text{ kg} \times (665-85) \times 100}{1.8 \text{ TPH} \times 1000 \text{ kg} \times 3200},$$

$$= 80.0\%$$

$$\text{Ratio of evaporation} = 8 \text{ TPH of steam}/1.8 \text{ TPH of coal}$$

$$= 4.4$$

### 8.1.3   ENERGY CONSERVATION OPPORTUNITIES IN BOILER

There are various opportunities to maintain the energy efficiency in boiler systems, which can be related to combustion, heat transfer, avoidable losses, high auxiliary power consumption, water quality and blow down. The following factors are examined in order to find out whether a boiler is being run to its maximum efficiency or not:

1. **Stack temperature**: The stack temperature ought to be as minimum as could be possible. Provided, it should not be below such that the exhaust of the water vapor is accumulated on the stack walls. It is essential in fills having critical sulfur content, having a very less temperature can cause dew point erosion of sulfur. Stack temperature with more than 200°C shows possible recollection of wasted heat. So, it is likewise that the scaling of exchange of recovery of heat and subsequently bringing of an early shut down of the cleaning of the water/flue side.

2. **Economizer for pre-heating the feed water**: Commonly, the flue gases coming from a modern 3-pass shell boiler have temperatures of 200°C–300°C. Subsequently, there is a possibility of recouping heat from these gasses. The temperature of the vent gas leaving the boiler is typically kept over at least 200°C, so the SOx in the flue does not result in condensation and erosion at surfaces of heat exchanger. At the point when a fuel of clean nature, for example, LPG or gas oil is utilized, the recovery heat economy should be worked out, as the temperature of the flue gas might be well beneath 200°C. The amount of savings related to energy relies on the boiler introduced and the fuel utilized. For a common model of shell boiler, having flue gas leaving at a temperature of 260°C, needs an economizer to lessen it to 200°C, increasing the temperature of the feed water by 15°C. Increment in general thermal efficiency is to be around 3%. For modern 3-pass shell boiler firing with a natural gas leaving at a temperature of 140°C with a condensing economizer can decrease the outgoing temperature to 65°C and increase thermal efficiency by 5%.

3. **Preheating of combustion air**: Preheating of combustion air is another option of feed water heating for waste heat recovery. With a specific goal to enhance thermal efficiency by 1%, the combustion air temperature is to be increased to 20°C. Most gas and oil burners utilized as a part of a boiler system are not used for preheating the air to high temperatures. Present-day burners take substantially high ignition preheating of air. Hence, it is conceivable to consider such devices as heat exchangers in exit of the flue gas as another option to an economizer when space or a high feed water return temperature making it practical.

4. **Incomplete combustion**: Incomplete combustion results from a lack of air or excess of fuel or improper fuel distribution. It is generally evident from color or smoke and should be rectified quickly. On account of system related to oil and gas firing, smoke or CO with normal or increase in abundance of air demonstrates issues in the burner system. A more recurring reason for incomplete combustion is the poor blending of fuel and air at the burner. Improper oil flames can due to inappropriate viscosity, worn tips, carbonization on tips, and deterioration of diffusers or spinner plates. With the firing of coal, unburnt carbon can lead to a significant loss. It happens as grit carryover or carbon-in-ash and can add up to over 2% of the supply of heat to the boiler. Fuel size being not uniform can be a reason for deficient burning.

In the case of chain grate stokers, huge lumps would not burn entirely, and little and fine chunk may hinder the air passage, consequently resulting in poor air circulation. In sprinkler stokers, stoker grate state, fuel merchants, over-fire systems etc., can influence the loss of carbon. Increment in the fines in pulverized coal additionally enhance loss of carbon.

5. **Excess control of air**: Surplus air is needed in every real-life case to guarantee complete combustion, to consider the typical varieties in burning and to guarantee certain level conditions for some fuels. The ideal excess air level for extremely efficient boiler happens for the sum of the losses because of incomplete combustion and loss due to heat in flue gases is minimum. Such level differs with furnace design, burner type, factors associated with fuel and processes. It can be found by conducting tests with various proportions of air-fuels.

   **Methods to control excess air:**
   - Oxygen analyzer of portable nature and draft gauges are used for recording periodic readings as a guidance for the user to control the flow of air manually to increase the effective working of the operation. Excess air lessening up to 20% is feasible.
   - Continuous oxygen analyzer is the most generally used method having readouts of locally mounted on draft gauge, due to which the user can control the flow of air. There is more decrement of 10%–15% that can be done on the system used previously.
   - The continuous oxygen analyzer can also has a pneumatic damper which is controlled by remote due to which readings are available in the control room. This helps the user to control various firing systems altogether remotely. The latest advanced technology is the automatic stack damper control, which is excellent for larger systems.

6. **Heat loss due to radiation and convection**: The outside boiler surface of shell type is hotter than its environment. The boundary walls hence lose heat to the outside environment relying upon the outer surface area and the temperature difference between the environment and the outer surface.

   The loss of thermal energy from the shell of a boiler is typically a loss of fixed value, independent of output of the boiler. With modern-day designs of boilers, it is only just 1.5% on the gross calorific value at the final rating, however, with an increment of about 6%, if the boiler works at just 25% output. Repairing or supplement insulation can diminish loss of thermal energy from the walls of the boiler and the pipes.

7. **Automatic blowdown control**: Constant blowdown in an uncontrolled manner is a colossal waste. Blowdown control of automatic nature is a potential device for assessing the water conductivity and pH of the boiler. A 10% blowdown in a 15 kg/cm$^2$ boiler results in the loss of efficiency of 3%.

8. **Reduction of boiler steam pressure**: This is a good process for decreasing the consumption of fuel, if allowed, by a possible of 1%–2%. Less pressure of steam gives a lower saturated steam temperature and without of stack heat recovery, a comparable decrease in flue gas temperature occurs.

## 8.2 INDUSTRIAL HEATING SYSTEM

**Furnaces**: A furnace is a device for melting metals for the purpose of casting, heat materials in order to change its shape (e.g., forging and rolling) or property such as annealing.

### 8.2.1 CLASSIFICATION OF FURNACES

Depending on the type of heat generation, the classification of furnaces is done on two basis, namely combustion type and electric type. The first one depends on the type of combustion and mainly distinguished and classified as oil-fired, or coal-fired or gas-fired.

- Depending on the type of material charging used in furnaces can be classified as:
  1. Intermittent or batch or periodical furnace
  2. Continuous furnace
- Depending on the type of the waste heat recovery, for example, regenerative and recuperative furnaces.
- Another classification is done on the basis of transfer of heat, type of charging, and type of recovery of heat.

#### 8.2.1.1 Oil Fired Furnace

Furnace oil is the significant fuel utilized as a part of oil-based furnaces, particularly to reheat and heat treatment of materials. The LDO is utilized as a part of furnaces where sulfur presence is unwanted. The maximum productivity of furnace operation lies in the complete combustion of fuel with the least excess air. Furnaces working on low efficiencies about 7%, however 90% is achievable in other combustion equipments, for example, boiler. It is deu to higher temperature working of furnace to take care of the demand. Such as, a stock being heated up to 1200°C will be having exhaust gasses going out at a minimum of 1200°C which follow a tremendous loss of heat via stack.

#### 8.2.1.2 Typical Furnace System

1. **Forging furnaces**: The forging furnace is utilized for preheating of billets and ingots to accomplish a "forge" temperature. The temperature of the furnace is kept at about 1200°C–1250°C. The forging furnace type, utilize an open type fireplace system and the significant part of the heat is disseminated by radiation. The typical loading in forging furnace is about 5–6 tons with working of 16 h or more every day. The aggregate working cycle can be split into (i) heating up time, (ii) soaking time, and (iii) forging time. Specific fuel consumption relies on the material type and the number of reheat required.

#### 8.2.1.3 Rerolling Mill Furnace

1. **Batch type**: A box type furnace is utilized in batch kind rerolling factory. The furnace is mainly utilized for scrap heating, little ingots and billets having a weight of 2–20 kg for rerolling. The material charging and discharging are manually done and the end products are rods, strips, etc. The working temperature is around 1200°C. The duration of the cyclic process is classified into heating up time and rerolling time. During heat-up, the material

is heated up to the prerequisite temperature and is removed to rerolling. The normal yield of furnace ranges from 10 to 15 tons/day and the specific fuel consumption differs from 180 to 280 kg of coal/ton of heated material.

2. **Continuous pusher type**: Procedure and working cycles of a continuous pusher type is similar to the batch type furnace. The working temperature is around 1250°C. For most cases, such furnaces work for more than 8 h with a yield of 20–25 tons everyday. The material or stock restores partial heat from flue gasses during its movement through the furnace length. Heat absorption in furnace for the material is slow, consistent, and constant all through the cross-section in contrast to batch kind.

   a. **Continuous steel reheating furnaces**: The principle purpose of a reheating furnace is to increase the steel temperature, generally in the range of 900°C–1250°C, up to the point when it is sufficiently having plastic nature to be squeezed/pressed/rolled, to the desired section, specific size, and shape. The furnace should likewise have particular requirement or targets as far as stock heating rates for productive and metallurgical causes. In consistent reheating procedure, the stock of steel forms a smooth material flow and is to be heated up to a preferred temperature as it passes in the furnace.

The furnace features are as shown in Figure 8.1.

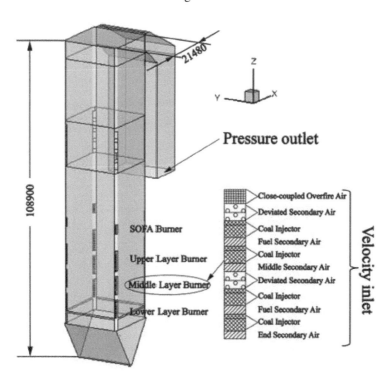

**FIGURE 8.1**   Furnace feature.

- A refractory chamber developed by using the materials of insulating type for holding heat at the high working temperatures.
- A hearth to support or convey for steel. This can comprise of materials of refractory type or a setting of metallic backings which could be water cooled.
- Burners that utilize gaseous/liquid fuels to increase and maintain the chamber temperature.

Reheating is done with coal or electricity. A process of removing exhaust gases of combustion from the chamber.

- A process of launching and removing the steel from the chamber.
- Such a process depends on capacity and furnace types, the processed steel, shape and size, and common layout of the rolling mill.
- The main parts are roller tables, conveyors, furnace pushers, charging machines, etc.

### 8.2.2 FURNACE HEAT TRANSFER

The critical process by which transfer of heat to the steel takes place in a reheating furnace is shown in Figure 8.2. The basic idea of transferring the heat to the stock by:

- Flame radiation, products of the hot combustion and, walls and roof of the furnace.
- The process of convection of hot gases traveling over the stock surface.

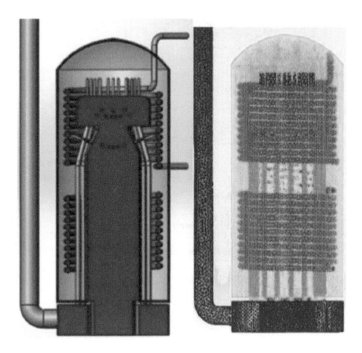

**FIGURE 8.2**   Heat transfer in furnace.

At a very high temperature, the most effective form of heat transfer is the radiation through walls in the reheating furnace. The transfer of heat due to radiation of the gas depends on the composition of the gas (concentration of $CO_2$ and water vapor), temperature distribution, and furnace geometry.

### 8.2.2.1 Types of Continuous Reheating Furnace

Continuous reheating furnaces are fundamentally classified on the basis of transportation of the stock via furnace. There are two principal routes:

- Stock are joined head to head to form a straight line after which it is forced into the furnace, these are called pusher type furnaces.
- Stock carries the steel via the furnace. It is kept on a movable hearth or supporting structure. This kind of stocks include walking beam, rotary hearth, continuous re-circulating bogie furnaces, etc.

The main parameters which are being considered related to the use of energy for furnace are the apertures of the inlet and outlet which should have the least size as possible and the designing should be done in such a way that to omit infiltration of air.

1. **Pusher type furnaces**: This type of furnace is widely used in the steel industries. The cost related to maintenance and installation is quite low compared to the moving hearth furnaces. The hearth of the furnace may be solid, but there is a possibility that the stock is pushed along the skids together with the support of cooling water allowing the stocks bottom and top surface to become heated. Pusher furnace design is represented in Figure 8.3.
2. **Walking hearth furnaces**: The walking heath furnace (Figure 8.4) allows step by step transportation of the stock through the furnace. These have several advantages, such as simple design and easily constructible, can handle stock of different sizes (within limits), the loss of energy by cooling water is lowest and minimal physical marking on the surface of the stock.

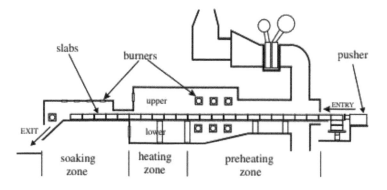

**FIGURE 8.3** Pusher type furnaces.

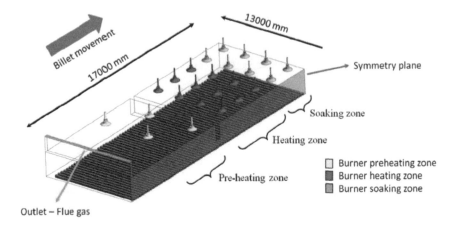

Outlet – Flue gas

**FIGURE 8.4**  Walking hearth type furnace.

The disadvantage of such type of furnace is the heating of the bottom surface of the stock is rare. It can be avoided by keeping vast spaces among the stock pieces. Small spaces among the single pieces of the stock do not allow the heating of the side faces and create the chances of a difference in temperature which is not required in the stock during discharge. Also, the stay time of stock is as long as several hours; which have unfavorable effects on the flexibility and yield due to scaling.

3. **Rotary hearth furnace**: The rotary hearth furnace (Figure 8.5) manages to replace the continuous re-circulating boogie type. The bogie's heating and cooling effects are removed, hence heat storage losses are very less.

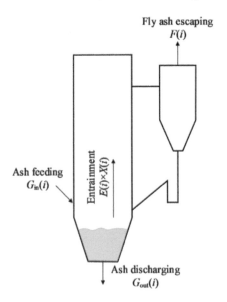

**FIGURE 8.5**  Principle of fluidization.

The rotary hearth design is very complicated in nature and have revolving and annular shaped.

4. **Continuous re-circulating Bogie type furnaces**: These are moving hearth kind furnaces have a tendency to be utilized for dense stock of flexible size and geometry. In bogie furnace, the stock is set on a bogie with a refractory hearth and goes along through the furnace with others in a long chain-like manner. The whole furnace length is continuously made of bogies. The bogie furnace tends to be long and strait and experience issues emerging from deficient fixing of the space between the bogies and furnace shell, scale removing troubles, and challenges related to firing over a tight hearth width.

5. **Walking beam furnace**: This type of furnace vanquish a large number of issues of pusher type furnaces and allows heating of the stock base. This permits lower stock heating periods and length of the furnace and hence improves to control heating rates, constant stock discharge temperatures, and operational adaptability. In a similar manner, as the pusher furnace having top and the bottom part, in any case, a significant segment of the furnace is underneath the mill level; this might be a limitation in a few applications.

## 8.2.3  FURNACE ENERGY SUPPLY

Since the by-products of flue gasses directly in contact with the stock, choosing the type of fuel is most significant. For instance, a few materials are not allow sulfur in the fuel. Likewise, utilization of solid fuels will lead to create small particulate matter, which tamper stock inside the furnace. Therefore, most of the furnace utilizes liquid fuel, gaseous fuel or electricity as energy input. Furnaces used for the purpose of melting steel or cast iron utilize electricity in induction and arc furnaces. Oil is a source of fuel for the melting of non-ferrous materials.

## 8.2.4  FURNACE PERFORMANCE EVALUATION

Thermal efficiency of processes heating systems/equipments, for example, furnaces, heaters, ovens, boilers and kiln is the ratio of heat transported to a material and heat provided to the heating systems/equipment. The idea behind the heating procedure is to supply a specific amount of thermal energy into a product/material increasing it to a specific temperature to set for further procedure or change its physical/mechanical properties. The material is heated in the furnace in order to achieve it. The outcome of this process is loss of energy in various zones and in various types. The whole process can be drawn in the Sankey chart. In all the heating equipments, huge quantity of the heat provided is wasted in exhaust gases.

The heat losses in furnace are:

• Storing of heat in the structure of the furnace.
• Heat loss from furnace structure or outside walls.
• The transport of heat to outside the furnace through conveyors, trays, etc.

- Loss due to radiation from various openings, hot exposed parts etc.
- Heat loss due to infiltration of cold air in the furnace.
- Heat loss due to excessive air supplied in the burner.

## 8.3  FLUIDIZED BED COMBUSTION (FBC) BOILERS

### 8.3.1  INTRODUCTION TO FBC BOILERS

There are a large number of limitations associated with the conventional grate fuel firing systems and hence, they are not economical for future use. FBC boilers come into existence as feasible substitute, which have a large number of advancements over the traditional boilers. The advantages of FBC are: compact boiler design, flexibility with the fuel used, higher combustion efficiency and reduced emissions of noxious pollutants ($SO_x$ and $NO_x$). The fuels used in such boilers are; coal, bagasse, rice husk, and other agricultural waste, etc. They have vast ranges of capacity from 0.5 to 100 T/h.

### 8.3.2  MECHANISM OF FLUIDIZED BED COMBUSTION

When a uniformly distributed air is moved upward through a finely divided bed of particles of sand supported on fine mesh then the particles are remain undisturbed at low velocities. With the slowly increase in air velocity, a state come when each and every particle remains suspended in the air stream and the bed is said to be fluidized. Now on more increase in the velocity, there is a formation of bubbles, turbulence and followed by mixing result the formation of dense surface beds. The solid particle bed behaves like a boiling fluid known as "bubbling fluidized beds." The bubbles disappear at higher air velocity and the particles are left out the bed. Therefore, recirculation of the particles is needed for maintaining a stable system and is known as "circulating fluidized beds" as represented in Figure 8.5.

The extent of fluidization is dependent on the size of the particles and the velocity of air. This means for a stable system, the extent of fluidization is less than that of gas. The difference between the mean velocity of solid and gas is the slip velocity. The slip velocity should be maximum for the solid and gas for excellent heat transfer and proper contact. Heating the particles of sand up to the ignition temperature of the fuels, and there is a constant injection of fuel, hence there will be a rapid burning of the fuel and attaining homogenous temperature throughout the bed. The FBC combustion takes place at 840°C–950°C. This temperature is much lesser than the ash fusion temperature and hence, problems related to melting of ash are avoided.

**Fixing, bubbling & fast fluidized beds**: The velocity of the gas passing via particle bed increases, reaches a value where the bed is fluidized and there is formation of bubbles as in case of boiling liquid. At high velocities, there is no such formation of bubbles and the solid are quickly moved out from bed and hence, it has to be recirculated to maintain the stability.

Combustion process is involved with 3 "T"s such as time, temperature, and turbulence. In FBC, fluidization encourages turbulence. Improved mixing results in excellent distribution of heating at a lower temperature. The residence time is very much higher than that of the conventional grate firing. Hence, in FBC system, the release of thermal energy is more efficient during the low temperature. Limestone is used in the bed for controlling the formation of $SO_x$ and $NO_x$ emission in the combustion chamber without any control system. This is one of the main benefits of FBC boilers.

## 8.4 COGENERATION

Cogeneration or combined heat and power (CHP) is said to be the series production of two dissimilar form of usable energy from one primary source, generally thermal energy and mechanical energy. The usage of mechanical energy is to run the alternator for the production of electrical energy or for the rotation of the equipment such as driving of the pumps, motor, compressor etc. Thermal energy is used directly in the process applications or indirectly in the form of steam, hot water production, heating air for drying, chilled water for process cooling, etc. Cogeneration has a wide range of applications in various areas and it's very much economical in nature. The overall efficiency of cogeneration is up to 85% and also more in some cases.

Followed by the reduction in the use of fossil fuels, cogeneration also benefits in the reduction of greenhouse gas emission. On-site production of the electricity reduces the pressure on utility and also losses associated with the transmission and distribution are avoided. Cogeneration is useful for both macro and micro point of view. In macro perspective, it distribute the financial burden among the government and the private sectors and also various local sources of energies are conserved in this process. At micro level, there is the reduction in the net bill since there is simultaneous on-site availability of power and heat and a reasonable energy tariff can be implemented throughout the country.

### 8.4.1 NEED FOR COGENERATION

The global primary source of electricity is the thermal power plants. The traditional method for power generation and supply to the customer is a waste ful process because effectively one-third of the primary energy is available to the end-user in the form of electricity and rest is actually wasted in the form of either heat loss or transmission and distribution losses.

The significant loss in the process of conversion is the rejection of heat to the surroundings water or air because of constraints of thermodynamic cycles used in power generation.

Also, 10%–15% loss is associated during the distribution and transmission of electricity through the electrical grid.

### 8.4.2 STEAM TURBINE COGENERATION SYSTEMS

The steam turbines are of two types, namely the backpressure and the extraction-condensing types.

The extraction-back pressure turbine is used in topping cycle cogeneration system when the end-user desires thermal energy at two different temperature stages. The full-condensing steam turbines are installed at sites where rejected heat by the process is used for power generation. The steam turbines have advantages as compared to other prime movers like widespread variation of using conventional and alternative fuels, for example, coal, natural gas, fuel oil, biomass, etc. Compromise up to some extend with power generation efficiency can be done for maintaining heat supply. Steam turbines are generally installed when the demand of electricity is more than one MW to a few hundred of MW. However, their operation is not appropriate for intermittent energy demand sites because of system inertia.

### 8.4.3  Gas Turbine Cogeneration Systems

Gas turbine cogeneration system has the ability to produce part or all energy requirement of the site and the heat energy is liberated at higher temperature in the exhaust stack and it can be recuperated for many heating and cooling utilities. The most common fuel is the natural gas but also other fuels (diesel, light fuel, etc.) are used. The range of the gas turbine varies from a fraction of MW to 100 MW.

The gas turbine cogeneration has the maximum growth in current times due to the larger accessibility of natural gas and the rapid development in technology, and a notable decrease in cost of installation, improved environmental performance etc. Moreover, the gestation period for gas turbine cogeneration venture is lesser and the equipments can be delivered on a partly basis. The starting operation time for the gas turbines is very less and the option of intermittent operation is available. The conversion efficiency heat to power is meager but a significant amount of heat can be recovered at high temperatures. In case of heat output is lesser than the needed value, additional natural gas firing is done via mixing extra fuel to the oxygen-rich exhaust gas for enhancing the thermal output.

### 8.4.4  Reciprocating Engine Cogeneration System

Reciprocating internal combustion (I. C.) engines-based cogeneration systems possess higher power generation efficiencies as compared to other prime movers. The heat recovery in this system has two sources—the high-temperature exhaust gas and the water cooling jacket system of the engine at low temperature. For smaller system, the heat recovery is very much efficient and is usually popular among energy consumption facilities that are smaller in size, generally, where electricity need is higher than the thermal energy needs and high-quality heat is not required.

### 8.4.5  Classification of Cogeneration Systems

The cogeneration systems are grouped into topping or bottoming cycle based on the sequent of energy use. In the topping cycle, the fuel supply is firstly used for power production followed by thermal energy (by-product) and thermal energy used for process heating and other thermal equipments. Topping cycle cogeneration has extensive applications and is very much well known cogeneration technique.

There are four kinds of topping cycle cogeneration systems and they are as follows:

1. Diesel engine or gas turbine generating mechanical or electrical power have heat recovery boiler to produce steam for driving secondary steam turbine, is known as combined cycle topping system.

2. The second category of topping cycle cogeneration system burns any kind of fuel to generate high-pressure steam which passes to a steam turbine to generate power, its exhaust provides process steam of low pressure. It is known as steam-turbine topping system.
3. In the third kind, heat is recovered from the exhaust of an engine or cooling jacket and flowing to heat recovery boiler for conversion into process steam for reuse.
4. In the fourth kind of gas-turbine topping system, natural gas burns in gas turbine for driving a generator. Then exhaust gas passes to heat recovery boiler to produce process steam and process.

In a bottoming cycle, the primary fuel is used to produce thermal energy at high temperature. The rejected heat from this process is further used to produce electrical power with the help of recovery boiler and a turbine generator. Bottoming cogeneration cycles are appropriate for product manufacturing industries. These industries need thermal energy at high temperature in furnaces and kilns, and also discard heat at enormously high temperatures. Bottoming cogeneration cycles are widely used in cement, steel, ceramic, gas, and petrochemical industries. The bottoming cycle plant is not much usual as compared to the topping cycle plant. In bottoming cycle, combustion of fuel takes place in a furnace to create synthetic rutile. Hot exhaust waste gases of furnace are employed in a boiler to produce steam that drives the turbine to produce electrical power.

## 8.5   BIOMASS UTILIZATION IN FBC AND CO-GENERATION TECHNOLOGY

This process used rice husk as fuel in a boiler of FBC for developing steam from medium to high pressure followed by the generation of electrical energy using steam turbine and some portion of the steam is also utilized in the manufacturing industry as process steam. In the previous sections, FBC boilers, steam and gas turbines, used for cogeneration are discussed for developing a understanding the equipment. There are many configurations possible but for the small- and medium-sized operations, there are only four and are as follows:

1. Generation of steam from FBC based boiler and no generation of electricity.
2. Generation of steam from FBC based boiler and electrical generation by the use of back-pressure type turbine.
3. Generation of steam from FBC based boiler and electrical generation by the use of extraction cum condensing type turbine.
4. Generation of steam from FBC based boiler and electricity generation by the use of a condensing type turbine with no involvement of steam in the process.

The fourth configuration hardly used for industrial application, it is more applicable for thermal power plants where fuel is biomass for the generation of steam from FBC based boiler. The system and the design configuration of the boilers and the turbines depends on where it is being used. It needs through feasibility study and survey in order to select the proper technology. Other than this, the selection is decided by accessibility of fuel and electricity, economic feasibility, etc. Industries hardly go with cogeneration systems if the charges and availability of electricity supply from the grids are adequate. In such cases, FBC boilers are installed to fulfill the process steam requirement (Configuration 1).

Steam quality, quantity, and pressure necessity of the industry decides the technology execution among configuration 1 and 2.

The process steps of biomass-based cogeneration system using FBC boiler depends on the site, nature and biomass quality, the local environmental regulations, etc.

## 8.6   WASTE HEAT RECOVERY

"Waste heat" is the heat generated  in the process of fuel combustion or chemical reaction which is discarded into the environment; however, it has the potential to reuse in some industrial process. The quality of heat is not good as of the previous but has some value. The process of recovering such heat is temperature dependent of the waste hot gas and economic feasibility.

Boilers, kilns, ovens, furnaces, etc., produces huge amount of hot flue gases. If few of the wasted heat is recycled, then a large amount of primary fuel will be saved; however, heat energy lost in flue gases are not fully recoverable.

### 8.6.1   BENEFITS OF WASTE HEAT RECOVERY

There are two significant benefits of waste heat recovery and are as follows:

#### 8.6.1.1   Direct Benefits

The waste recovery process effects on efficiency of whole system/process, which is represented by the lessening of utility, cost, and the cost related to process.

#### 8.6.1.2   Indirect Benefits

1. **Reduction in pollution**: Many toxic combustible elements/gases like carbon monoxide, sour gas, carbon black off gases, oil sludge, acrylonitrile, and other plastic chemicals, etc., discharge to the atmosphere when waste is burnt in the incinerators. Waste heat recovery technology serves dual purpose, i.e., recovery of useful heat and reduces the impact on environment.
2. **Reduction in equipment sizes**: Waste heat recovery process reduces consumption of fuel, hence the reduction of flue gas also leads to reduction in the size of flue gas handling equipments and machinery.
3. **Reduction in auxiliary energy consumption**: The size of the equipment/ parts and machinery is reduced. It results reduction in the consumption of auxiliary energy systems such as fans and pumps, etc.

## 8.6.2   Development of a Waste Heat Recovery System

Understanding of proper concept is necessary for developing the system of waste heat recovery. This is possible by continuous review of the process floor sheets, diagrams of layouts, piping isometrics, electrical, and instruments wiring, ducting etc. The detailed study will support in recognizing:

1. The source of wasted heat and where it can be used.
2. Problems in the plant due to recovery of waste heat.
3. Availability of space.
4. All other parameters which are acting as a constraint in the development of waste heat recovery system. For example, dew formation starts in the equipment, etc.

The economic analysis (investment, depreciation, payback period, rate of return, etc.) of selected waste heat recovery system is essential for taking rational decision. Once, waste heat source and practicable application of it are recognized, the following identification of appropriate heat recovery equipment and system should be retrieved.

### 8.6.2.1   Commercial Waste Heat Recovery Devices

#### 8.6.2.1.1   Recuperators

Recuperator is a heat exchanger where heat exchange process occurs between the hot flue gases and air, with the help of metallic or ceramic wall medium. The air needed for the combustion is carried by ducts or tubes for preheating in the opposite side that contains a stream of waste heat.

The most accessible design of a recuperator is the metallic radiation recuperator, consisting of two concentric metal tubes, hot exhaust gas is carried by the inner tube and the external carries combustion air to the furnace burner inlet from atmosphere. Inlet combustion air cools the hot flue gases by which it transports supplementary energy into the combustion chamber. This added heat energy is independent of the fuel supply and hence, less fuel is needed in the furnace.

#### 8.6.2.1.2   Regenerator

The process of regeneration is done where the capacity is very high, for example, glass and steel melting furnace. There is a well connection between the regenerator size, time during reversals, brick thickness, brick conductivity, and ratio of the heat storage brick.

Reversal time is a critical parameter in the regenerator. Large time duration increases thermal storage which resulting in high cost. Reversal large periods consequence the lower mean temperature of preheat, therefore reducing the fuel economy.

#### 8.6.2.1.3   Economizer

For a boiler system, economizer uses the heat of the hot flue gas in the preheating of boiler feed water. In case of air preheater, waste heat is used to heat inlet combustion air. In both the circumstances, there is a reduction in the amount of the fuel used in the boiler. There is 1% saving in boiler fuel either by increasing every 6°C in feed water temperature using economizer or 20°C escalation in combustion air temperature via air preheater.

### 8.6.2.1.4   Shell and Tube Heat Exchanger

If liquid or vapor is the media of having waste heat and also heat transfer to liquid, in such cases shell and tube heat exchangers are employed as both paths should be sealed to sustain high pressures of their respective fluids. The shell holds the bundle of tube with internal baffles. Baffles guide the direction to fluid in the shell over the tubes in multiple passes manner. The strength of the shell is comparatively weaker as compared to the tubes since high-pressure fluid is passed through the tubes and the low-pressure fluid is passed through the shell. The vapor containing the waste heat condensed because transfer of latent heat to the liquid is being heated.

### 8.6.2.1.5   Plate Heat Exchanger

The cost of heat exchange surfaces is one of the important economic consideration when the temperature differences between fluids are not high. Such kind of problem can be overcome by using plate type heat exchanger. It has series of separate parallel plates arrangement making thin flow pass. The plates are separated from the adjacent plates by using gaskets and the steam is passed parallelly in the alternative plates while fluid to be heated passes in between the hot plates. Corrugated plate can be used in order to increase the rate of heat transfer. The hot liquid is passed through the bottom port and is allowed to move upward among every alternate plate while the liquid of cold nature is at the top of the head and is passed downwards alternatively. When the direction of the fluids are opposite, such configuration is called counter current. Plate heat exchangers are extensively used in pharmaceutical plant, food industries, refineries, etc.

### 8.6.2.1.6   Thermocompressor

In various cases, steam at very low pressure are reused like water after condensation due to scarcity of good alternative reuse option. Sometimes, it becomes viable to compress low pressure steam via very high pressure steam and reuse like medium pressure steam. Latent heat is in large portion of energy in steam and therefore, thermocompressing renders huge reformation in waste heat recovery.

Thermocompressor is a modest equipment attached with a nozzle to accelerate high pressure (HP) steam into a high velocity fluid. The low pressure (LP) steam enters through momentum transfer, it further recompresses in a divergent venturi. Thermocompressor is employed in evaporators where the boiling steam is recompressed and used as heating steam.

## 8.7   THERMAL INSULATION

Thermal insulators are bad conductors of heat as it has low thermal conductivity. Insulation is done in buildings and process industries to avoid heat loss or heat gain. Though the primary objective of insulation is economic aspect; however, it also provides more precise control of process temperatures and personnel safety. It inhibits condensation over cold surfaces which results in corrosion. Insulating materials are porous and have huge amount of dormant air cells.

Benefits of thermal insulation are:

- Reduction in the net consumption of energy.
- Better control on process through commanding process temperature.
- Does not allow corrosion by keeping the refrigerated surface above the due point.
- Keeps the equipment fireproof.
- Vibration absorption.

## QUESTIONS

Q.1. Discuss the sources of heat wastage and its potential applications.

Q.2. With a neat sketch explain gas turbine cogeneration plant.

Q.3. What is cogeneration? State its necessity.

Q.4. What are cogeneration plants? Explain the difference between bottoming and topping cycle cogeneration plants.

Q.5. Explain in brief energy efficiency versus energy conservation. Write stepwise procedure to calculate boiler efficiency.

# 9 Building Energy Management

## 9.1 INTRODUCTION

The design of buildings is not a subject of study in India due to its poor thermal and electricity capabilities. The configuration of a buildings' uniqueness, development cost, etc., are the various parameters, which lead to the creation of building design. The building's operating energy cost has increased enormously as the energy efficiency factors were either not considered or ignored at the design stage.

The weather conditions of place denote the conditions of the atmosphere for a particular period. The entire weather conditions took over a year, and it was said to be the climate of that area of "macro-climate." These climate analyses can explain the situations occurring throughout the year, how much can a person experience the degree of hot or cold and find out the comfort and discomfort periods. Such analysis helps a designer of a building to design so that the adverse climatic conditions can be avoided as much as possible and make it as habitable and comfortable as possible. The discomfort occurring to a person as the usage of mechanical devices to reduce the discomfort can be abruptly reduced by taking advantage of the climatic effects. The building design and the outlets of the buildings can be done on such a basis. For example, in a place where the temperature is scorching and humid due to excessive solar irradiation (a), the amount of solar irradiation entering the building is to be reduced and (b), the building should be adequately ventilated to avoid outside weather changes affect the inside of the buildings.

The law of conservation of energy depends on minimization of energy waste and bridges the gap between the demand and the supply of power. For this purpose, the Bureau of Energy Efficiency was created and it is working under the Power Minister by the government.

One of the main points of the conservation of energy act was to imply energy conservation codes for building for causing conservation of efficiency of buildings.

The E.C.B.C. sets a barrier for the maximum energy consumption for the building. In case of the commercial buildings, the E.C.B.C. is considered widely due to programs formulated by the government that helps in finding the possibilities for savings of energy. It is to be noted that the building should be designed and built, taking into account the various energy efficiency criteria, which have been implemented since the beginning. Hence, creating codes for conservation in buildings is required for this purpose, the motive of the E.C.B.C. is to develop minimum barrier for the conservation of energy in buildings and create awareness for the consumption of energy.

Codes on energy conservation of buildings are reducing energy consumptions and saving the energy of buildings for a long time, while E.C.B.C. is proven to be beneficial in decreasing energy consumption in buildings. Just like the ASHRAE standard is followed in the United States, which manages to reduce the energy consumption of buildings by 60%.

For the developing countries, the reduction in energy consumption is about 20%–35% for buildings of the first-generation, and such savings are very much useful as the commercial buildings have 25%–35% of the electricity use of the country.

## 9.2   FACTORS AFFECTING CLIMATE

Both weather and climate are defined as climatic factors. The variables associated are as follows:

1. Amount of radiation of the sun
2. Temperature of the surrounding
3. Atmospheric humidity
4. Rainfall
5. Wind
6. Condition of the sky

1. **Solar radiation**: It is the radiation energy received by the sun. It is represented by the intensity of the sunrays falling per unit time per unit area as in watts per square meter ($W/m^2$). The amount of radiation on the surface is not constant. However, it depends on the geographical location (Figure 9.1a–c). The intensity and amount of solar radiation only determines the temperature of that area, whether it is hot or cold. The amount of radiation is measured with the help of the Pyranometer and Pyrheliometer, and the sunshine recorder measures the sunshine.
2. **Ambient temperature**: The temperature of a well-ventilated closed shaded area is known as the ambient temperature, represented in °C. The temperature of an area depends on various factors such as wind, radiation of the sun, water bodies, etc. During low-speed wind conditions, the local parameters tend to affect the temperature of an area. However, with more wind speed, the effect due to local parameters tends to decrease gradually. The results of various factors on the ambient temperature are represented in Figure 9.2a–c. A regular thermometer in the Stevenson screen can measure the ambient temperature.
3. **Air humidity**: The air humidity represents the amount of moisture present in the air and it is generally represented as "Relative humidity." The relative humidity is represented as the water vapor of certain mass to that of the same quantity of the saturated air volume at the same temperature, generally shown in percentage. It has a certain degree of variation starting from the dawn when it is maximum, and the air temperature in minimum and slowly decreases due to an increase in the air temperature. The relative

humidity is known to decrease to a maximum level during the summertime. Areas with a high level of relative humidity are tending to have low solar radiations due to absorption and scattering effect. High humidity decreases the water evaporation rate and increases the sweat. If there is both high humidity and temperature, there is irritation followed by it; the relations are shown in Figure 9.3a–c.

4. **Precipitation**: Precipitation can be of dew, rain or hail, represented in millimeters (mm) by a rain gauge. The precipitation effects in buildings are illustrated in Figure 9.4.

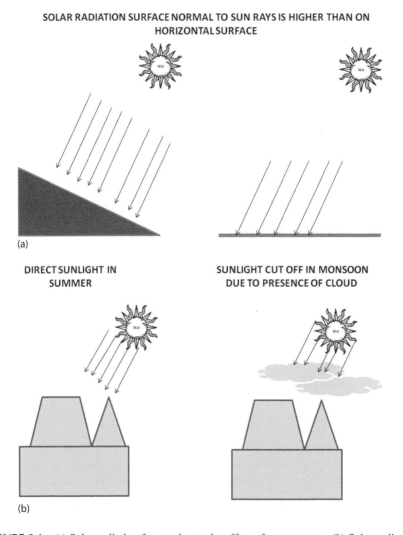

**FIGURE 9.1**   (a) Solar radiation factors due to the effect of arrangement, (b) Solar radiation factors due to covering of the sky.                                        (*Continued*)

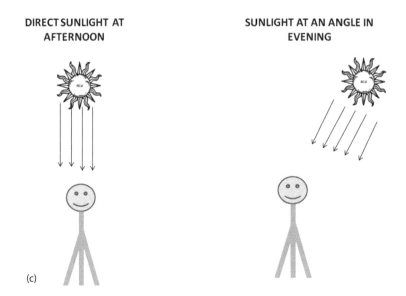

**FIGURE 9.1 (Continued)**   (c) Factors affecting solar radiation effect of time.

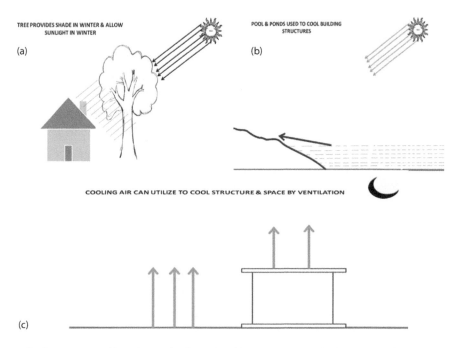

**FIGURE 9.2**   (a) Effect due to shading, (b) Effect due to water body, and (c) Effect due to sky conditions.

AIR MOVEMENT BY CROSS VENTILATION
CAN REDUCE DISCOMFORT

EVAPORATIVE COOLING CAN
PROVIDE COMFORT

(a)

CONDENSATION MAY LEAD TO DETERIORATION OF BUILDING
MATERIAL

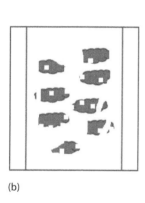

(b)

**FIGURE 9.3**   (a) Effects of high temperature and high humidity and high temperature and low humidity, and (b) Effects of low temperature and high humidity.

5. **Wind**: The movement of air due to changes or differences in heating effects of the earth's mass (both land and water), sun radiation, and the rotation of the earth is called wind. It is measured by a device called anemometer expressed in meters per second (m/sec). It is a significant parameter during the designing of a building since it affects the indoor comfort of a building due to the influence of convection heat current and as well as creating infiltration of air (Figure 9.5).

6. **Sky condition**: The cover of the clouds across during the sunshine is known as sky cover. During clear sky, solar radiation is much more than the cloud cover during the monsoon period. The loss due to re-radiation increases

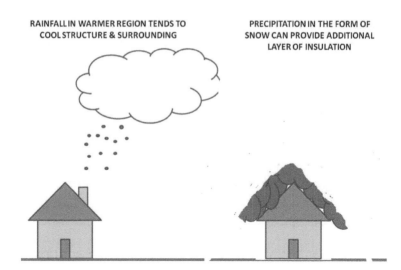

RAINFALL IN WARMER REGION TENDS TO
COOL STRUCTURE & SURROUNDING

PRECIPITATION IN THE FORM OF
SNOW CAN PROVIDE ADDITIONAL
LAYER OF INSULATION

**FIGURE 9.4**    Effects of rainfall and the effect of snow.

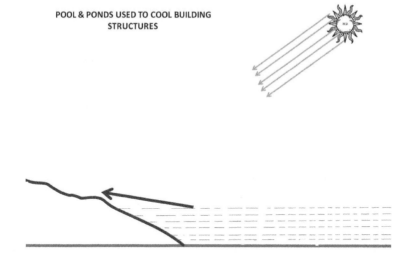

POOL & PONDS USED TO COOL BUILDING
STRUCTURES

**FIGURE 9.5**    Factors affecting wind.

during the clear sky condition and decreases during the cloud cover condition, as shown in Figure 9.6. Okta is the unit to measure sky cover. Example 3 states that 3/8 of the skies is covered with clouds.

Moreover, other parameters such as the contour; water bodies, vegetation, etc., also affect the local climate. The anthropogenic impact on the climate is due to human-made structures and pollution.

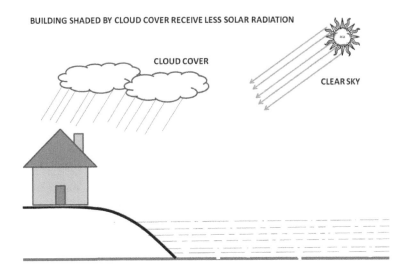

**FIGURE 9.6**   Effect of sky condition.

## 9.3   ENERGY CONSERVATION BUILDING CODE-PROVISIONS IN THE EC ACT 2001

As per the EC Act, "energy conservation building codes" states the rules and the regulations of consumption of energy denoted in terms of per square meter of the area where there is a usage of energy and the building location is taken into consideration.

The EC enables the Government under section 14(p) to carry out and illustrate the Building Code for energy conservation. Bureau of Energy Efficiency (BEE) has the duty in section 13(d) to take measures for preparing the E.C.B.C. guidelines.

The EC Act allows the E.C.B.C. for buildings having a load of 500 kW or contract demand of 600 kVA or more and are to be used for commercial use and should be built based on the E.C.B.C. rules as given by the Governments of the State under section 15(a).

The E.C.B.C. has the ability to change for current buildings. Such changes for the current buildings are applicable for exclusions and setting of riders, as done by E.C.B.C.

### 9.3.1   E.C.B.C. DEVELOPMENT

The method selected from developing the E.C.B.C. was to allow participants of all kinds who were involved in the development process, such as engineers, architects, other non-government institutions, etc.

The BEE comprised of a committee of experts (C.O.E.) and stakeholders that controlled such in-depth consultancy for three years. Such groups helped in forming appropriate requirements for designs and construction for a locality.

### 9.3.2 Broad Stakeholder Participation

The E.C.B.C. has been created to take control of the country's five different climatic zones, as for a case national building code. These climatic regions are composite, hot and dry, warm and humid, and moderate and cold.

It was taken into notice so that the building codes for energy regulation are easy to follow and use. The document of the Code has only to denote the requirement in performance in a format that is easy to use.

### 9.3.3 Features of E.C.B.C.

The E.C.B.C. will set a minimum efficiency standard for the non-commercial buildings for designing and construction purposes. The standards of performance for the energy of the current systems in the buildings are illustrated by E.C.B.C.

- Building envelope,
- Lighting,
- Heating, ventilation and air conditioning (HVAC),
- Heating of service water, and
- Electricity and its distribution.

However, the Code as required for the building envelope is the same for new as well as for extensions and modifications.

The provisions of E.C.B.C. improve the comfort and tenant profitability. The E.C.B.C. enhances energy effective plan or retrofit of business structures, so they are composed in a way that diminishes the utilization of energy without influencing the building capacity, comfort, well-being, or the efficiency of the tenants and with proper respect for financial contemplations. These codes remove the building design having high-energy demand and its related cost. Energy cost savings result from energy code compliance and it directly benefits the building owners and occupants over the life cycle of the building.

Everywhere throughout the world, E.C.B.C.s have a confirmed record of accomplishment of substantially decreasing the energy use in buildings cost effectively. The E.C.B.C. is cost effective for both; people and an individual building owner. The advantages of property proprietors incorporate decreased costs related to energy and enhanced comfort level. The advantages to society are:

- Reduction in investments related to supply of energy set up.
- Reduction in impacts in the environment.
- Improvement in the electrical reliability.
- Efficient using of resources.

E.C.B.C.

- E.C.B.C. takes care of the use of energy based on unit area of the floor and other high energy using systems like HVAC, lighting.

- E.C.B.C. illustrates the parameters in detail of the different usage of material for the building to increase the net performance of building envelope by reducing the gain of heat and loss due to cooling.
- E.C.B.C. mentions the use of optimized glass combinations for increasing daylight and reducing the load due to cooling.

The E.C.B.C. compliant an estimated potential saving of around 30%–35% for each building.

### 9.3.4   E.C.B.C.—Major Elements of the Building Plan

The E.C.B.C. looks for need of performance evaluation. However, they follow specific criteria, which include various data and details of the materials and frameworks of the building in a detailed manner are as follows:

- **Building Envelope**: Materials for insulation, fenestration, solar heat gain coefficients (SHGC), and so on.
- **Heating, Ventilation, and Air Conditioning**: System and equipment types, sizes, and efficiencies..
- **Service Hot Water and Pumping**: Heat of water using solar power.
- **Lighting:** Lighting schedule showing type, number, and wattage of lamps and ballasts.
- **Electrical Power**: A chart showing the loss of electrical data and the power lost due to the efficiencies of the transformer.

### 9.3.5   E.C.B.C. Benefits

In India, studies recommend that around 20%–25% of the total demand of energy is because of manufacturing materials required in the areas of the construction, while another 15% is required in order to run the building.

The proposed Energy Conservation Building Code sets a base proficiency standard for business building in all climatic condition in the nation. The calculated decrease in the use of energy for new buildings was found between 25% and 40% subjected to the type of building, running hours, and climatic condition.

The advantages to society incorporate diminished capital interests in the supply of energy infra, decreased ecological effects, and more efficient utilization of resources.

The current energy conservation building code has a minimum standard efficiency for all buildings of commercial types based on climatic zones of the countries. The calculated deduction for the new buildings is about 25%–40%.

The average use of energy for buildings of commercial type is 200 kWh/mr$^2$/year. Strict application of E C.B.C. can lead to the reduction of usage of energy to about 30%–40% to 120–160 kWh/m$^2$/year. In 2004–2005, construction of resident was taken to be 19.25 million m$^2$, and construction of commercial is 21.50 million m$^2$. In 2005–2006, there was an expected increase of 10% based on the previous year.

The strict application of E.C.B.C. was expected to have an output savings of 1.7 million units in the initial year when it was implemented based on data.

### 9.3.6  E.C.B.C. Implementation—All India Level

The purpose of the E.C.B.C. is for ensuring:

- The process of compliance is simple, easy, and clear to understand.
- The techniques and methods are to be used globally for India's benefit.
- The path of compliance made to benefit the major stakeholders.

Strict measures are to be taken by the government for ensuring proper supply of information regarding E.C.B.C. application proceedings and making it friendly to the consumer. Our commitment is also made for identifying proper methods of encouragement of compliance, including awards of monetary types, awards of publicity, developing labels of energy, and assistance of technicality.

### 9.3.7  Road Map to Make E.C.B.C. Mandatory

There is an inadequate capacity in a country for implementing the Code of E.C.B.C. and hence, the implementation of the codes will be based on volunteers. Incentives will be provided during this phase of voluntary and also promotion in the industry for insulators, windows, etc. When there will be sufficient availability of expertise of technical level and material of compliance only then the codes are to be made mandatory. Furthermore, the awareness campaign by the government is to be conducted throughout the country for the promotion of E.C.B.C.

## 9.4  ENERGY CONSERVATION MEASURES

### 9.4.1  Building Envelope

For a few buildings, the building cover (i.e., wall, roof, etc.) has an important role to play on the usage of energy for conditioning. The energy efficiency can be found out by its building load coefficient (B.L.C.). It can be found by the analysis of regression of the utilization data or by calculation of the resistance offered by the materials of construction used in the building cover. Some of the common techniques used for the improvement of the performance of the building cover as follows:

1. **Addition of thermal insulation**: Building surface having no thermal insulation has very much less cost associated.
2. **Replacing the windows**: Windows occupy a large amount of portion in the buildings; using proper energy efficient windows can help in the reduction of uses of energy and can improve conditions.
3. **Reducing the leakage of air**: The area of leakage in a building cover can be reduced by using techniques of weather-stripping. In buildings of residential purposes, the rate of infiltration can be found out by a test setup of a blower door.

### 9.4.2  Ventilation and Indoor Air Quality

1. **Ventilation in Commercial/Institutional Buildings**: The energy needed to regulate air ventilation is very much useful for both buildings of commercial

and industrial types, which is generally applicable in locations of harsh conditions of weather. The purpose of ventilation is not only to provide fresh air to the building occupants but also to take control of the dust particles in the industrial area. During an audit, an estimation of the volume of fresh air and comparing it with the calculated amount of air from ventilation followed by how much is required based on standard and codes should be taken into account. If the ventilation is excess, then there should be a reduction; otherwise, it can lead to high heating and cooling. Moreover, in few climates at times during the day or year, ventilation can help significantly to reduce the heating and cooling effects using airside economizer cycles. If the ventilation of air is excess, a damper can be adjusted to control the ventilation based on requirements. More significant reduction of the air outside can be made by using ventilation controls supplying outside here only when needed. Generally, the following are essential requirements for effective control of ventilation:

- Unpredictable variations in the occupancy patterns,
- The requirement of either heating or cooling for most of the year, and
- Low pollutant emissions from non-occupant sources (i.e., furniture, equipment, etc.).

2. **Ventilation of Parking Garages**: The parking garage for an automobile can be partly open or entirely closed. The garages of partly open types having open sides above grades do not need any ventilation. The parking garages, which are entirely closed, are generally underground in the ventilation of mechanical types and needed. Ventilation of the parking areas faces problems related to the quality of air. The most severe problem is the emission of a high amount of carbon monoxide (CO) in the parking garages. Other problems related to the close garage are fumes from oil and other gases such as oxides of nitrogen ($NO_x$) and others from fuel products. For the determination of proper ventilation for garages, two factors are taken into account: forcibly the number of cars present and emission level. Next is the number of cars working depending on the type of service provided by the parking garage and it widely varies from 3% to 20% of the total vehicle capacity.

### 9.4.3 ELECTRICAL SYSTEMS

In commercial buildings and the industrial uses to electricity dominate the utility bill. Office equipment, motors, and lighting devices are the various devices of electrical nature, which consume most of the energy in a building.

1. **Lighting**: Generally, a building used for office has 40% usage of electricity by the lights. There are various methods of inexpensive types for increasing lighting efficiency. Such methods are energy efficient lights, using devices that are more reflective, de-lamping, and maximum uses of daylight.

2. **Daylighting**: Proper use of daylight is a very much cost-reducing and alternative option for electrical lights in buildings of commercial and institutional nature. Using automatic controllers and devices, daylight can help in the elimination of electrical light and help insufficient illumination of the office area.

3. **Office Equipment**: The equipment used in the office uses the maximum load of electricity, generally in the buildings of commercial nature. The office equipments are computers, fax machine, printers, photocopy machines, etc. Nowadays, a large number of manufacturers supply with office equipment of high-energy efficiency. For example, computer going into sleep mode when not in use.

4. **Motors**: The cost of energy in the operation of motors of electrical nature is an integral part of the budget of a building of industrial and commercial nature. Steps are taken for deduction in the energy cost related to the motor usage includes operation time reduction, optimizing the system of the motor, and using a particular controlling device for the optimum output based on demand by variable speed drives of air or water distribution and installation of energy efficient motors. Moreover, other than the total reduction in electrical energy usage, such types of electrical devices decrease the need for cooling and hence, reduce the use of electrical energy in the building. This reduction in the cooling energy and increment usage of thermal energy is taken into account during the evaluation of methods of reducing the cost and improving lighting systems.

5. **H.V.A.C. A.C. Systems**: The energy use in H.V.A.C. A.C. systems can represent 40% of the energy consumed by the building of commercial type. There are various measures, which can be taken for improving the performance of energy of the primary and the secondary H.V.A.C. A.C. systems. Few of them are illustrated below:

   a. Setting up temperatures of the thermostat. When suitable, set back of heating temperatures can be used during periods of unoccupancy. In addition, setting up of temperature for cooling can be done.

   b. Retrofit of constant air/volume systems. For buildings of commercial nature, variable air volume systems should be taken into consideration when the H.V.A.C. A.C. system is relying on the fans of constant volume types for regulating a part of the entire building.

   c. Retrofit of central cooling plants. At present, various chillers are very much energy efficient, and very easy to operate and handling and are useful for projects related to retrofit type. Usually, the chillers are very much cost-effective and energy efficient.

   d. Retrofit of central heating plants. The boiler efficiency can be improved very much by proper air-fuel ratio for optimum combustion. Moreover, there are new boilers in the market of energy efficient nature that are justifiable in terms of economy during the substitution of the old boilers.

   e. Installation of heat recovery systems. The H.V.A.C. A.C. systems can recover some of the heat loss.

   It is to be kept in mind that there is a strong interaction between the heating and the cooling systems. Hence, the entire analysis is to be done during the retrofitting of an H.V.A.C. A.C. system. Optimizing the use of energy for a central cooling plant is one of the methods of using the net systems method for a reduction in the use of energy in heating or cooling of buildings.

6. **Compressed-Air Systems**: Compressed air is a vital tool for various manufacturing processes. There is a wide range of instruments, which uses compressed air. Unfortunately, a large amount of compressed air is wasted in various processes. Calculation suggests 20%–25% of the energy input as electricity is used for the generation of compressed air. There is 10%–50% leakage of waste and 5%–40% of misuse of the compressed air. Various types of compressors are available such as centrifugal reciprocating or rotary, with one or multiple stages. The screw compressor is very much in use in the industries for units of small and medium-size. Table 9.1 gives the pressure rate, rate of airflow,

## TABLE 9.1
### Energy Saving due to Implementation of Various Passive Features

| Opportunities | Measures |
| --- | --- |
| Proper orientation (change) in longer axis from N–S to E–W. | 30% reduction in solar radiation incident on walls. |
| Proper shading of windows. | Over 10% reduction in cooling load. |
| Painting of external surfaces of walls with minimum absorption of solar radiation and high emission in longer wave region. | 40%–50% saving in electrical energy. |
| Application of high albedo coating on roofs. | 10%–43% saving in energy used to cool the buildings. |
| Change in roof albedo from 0.18 to 0.81. | 69% saving in cooling energy. |
| Shading of houses by nearby trees. | 30% reduction in heat. |
| Shading of walls by a row of trees. | 50% reduction in heat-gain through the wall. |
| Growing a thick layer of vines on wall. | 75% reduction in heat gain through the wall. |
| Covering the roof top with deciduous plants or creepers. | Roof top temperature is reduced by 15°C. Heat flow of 200 w/m$^2$ enters through uncovered roof, whereas 10 w/m$^2$ is transferred upward from inside the room. |
| Evaporative cooling of roof. | Cooling load is reduced by 40%. |
| Replacement of air within glazing by argon, krypton or xenon. | U-value is reduced to 1.3 and 1 w/m$^2$k, respectively. |
| Provision of cross ventilation. | Contributes significantly toward ameliorating thermal environment indoors. Saving in energy varies from case to case. |
| Nocturnal ventilation | In low mass building indoor maxima is very close to outdoor maxima. However, in high mass building, the indoor maxima is reduced by 3.5°C. |
| Radiative cooling using metallic nocturnal radiator. | The indoor air temperature is reduced by 4°C–6°C. |
| Earth air tunnel | The air is adequately cooled by 3°C as it passes through the earth coupled heat exchanger tube. |
| Use of renewable energy sources such as wind, solar, etc. | Saving in energy varies from case to case. |
| Use of hollow clay tiles for roofs. | Saving of 18%–30% of cooling energy. |

and the power needed for various compressors. The energy conservation techniques that can be employed for compressed air system are as follows:

a.  Repairing of the leak in the distribution line. There are various methods associated with the detection of leaks starting from water and soap to the usage of high and equipment example, ultrasound leak detectors.

b.  Reducing the temperature of the air at the inlet and increasing the pressure of air at the inlet.

c.  Reducing the usage of compressed air and pressure by modifying the process.

d.  Using systems of heat recovery for using the heat of compression in the process for water heating or heating of the building area.

e.  Installing automatic sensors for optimizing the compression operations and reducing load operations.

f.  Using a compressor booster in order to obtain Hyatt discharge of pressure. The booster for compressors is economical if the high pressure associated with the year is a part of the net compressed air used in the process. Without the compressor booster, the compressor of the primary standard would have to compress the whole year to the desired maximum pressure.

## 9.5  COMMERCIAL AND INDUSTRIAL BUILDINGS

### 9.5.1  BUILDING-ENVELOPE TECHNOLOGIES

Development of new materials and many systems have introduced to increase the efficiency of buildings, especially windows, which include:

1.  A special type of glas optimizing solar grains and effects of a shade.
2.  Chromogenic glazing changes their properties by itself based on temperature and light conditions.
3.  Using PV panels have the ability for electricity generation and observing solar energy at the same time resulting in the reduction of gain of heat from the building.

#### 9.5.1.1  Light-Pipe Technologies

Although the best method is daylight for areas close to the windows, but it is not very much useful for the interior parts. In modern technologies, it allows the piping of light from also rooftop or partition or wall collectors to inner spaces where natural sunlight is not available.

#### 9.5.1.2  HVAC Systems and Controls

The various strategies of energy retrofits are as follows:

1.  Comforts that can control heating and cooling in a building can reduce energy consumption. H.V.A.C. A.C. manufacturers have found various methods in tackling thermal comfort rather than just focusing on the temperature and is focused on developing sensors related to thermal comfort.

2. Technologies for heat recovery such as rotary heat wheels and heat pipes have the ability to recover 50%–80% of energy from hot or cool air ventilation in the building supply.
3. Dry cooling technology is nowadays available and can be used in buildings.
4. Heat pump of geothermal natural also helps in recovering heat from the underground for building usage.
5. Thermal energy storage offers cheap energy source to produce heating or cooling effect for optimizing the building conditions during the required period.

### 9.5.1.3 Cogeneration

The technology of cogeneration was never new, but recent developments in its thermal and electrical system had made the way in significant usage in various buildings of the sectors in our society.

## 9.6 ENERGY SAVING IN BUILDINGS DUE TO VARIOUS PASSIVE SYSTEMS

The studies carried out by various organizations on different passive features incorporated in buildings revealed that a considerable amount of energy might be saved. A few examples are reported in Table 9.1.

## 9.7 BARRIERS IN ADOPTING ENERGY EFFICIENCY IN RESIDENTIAL BUILDINGS

1. The current energy scenarios, the architect is not very knowledgeable about the energy efficient designs and its features. Awareness and technical skills are needed for the energy efficient design of a building.
2. The various features related to the efficiency of energy can only be applied by the client who is willing to pay the extra money for reducing the area of the flooring. The client should be aware about the need for efficient buildings.
3. The high-income people's lifestyle does not match with the conditions of energy efficiency. For them, energy consumption does not matter; it is only a status or a symbol.
4. The owners of a flat are not always living in their flat they are buying. It is given out as rent to tenants. Hence, the primary owner is not bothered about the energy efficiency and hence, does not bear any cost related to it.

It was seen using some of the primitive methods in designing leads to significant cooling and less use of energy, creating the inside environment very comfortable. It is also observed that despite the clinching shortage of conventional energy and well-known benefits of the use of non-conventional energies like solar and wind energy, the concept of passive systems could not find more extensive application in the design of buildings. Further, the basic principles of various passive systems are well defined. However, the technology of passive cooling is yet to reach the level of an established practice. The response of users of buildings provided with passive cooling systems to the private queries about the viability of the system also lacks consistency. Moreover, efforts are continuously being made across the glove to make further advancement in knowledge in the area of passive design of buildings. In this scenario, a hybrid passive system emerges a viable option that may provide greater reliability and attract the broader application of passive techniques in the design of buildings in the hot climate. Hence, hybrid passive cooling is an essential critical R&D area for functional and energy efficient design of buildings.

## QUESTIONS

Q.1. What are the various factors affecting the climate?

Q.2. Explain the energy conservation building code.

Q.3. Explain the various features of E.C.B.C.

Q.4. Explain the energy-saving opportunities in buildings due to various passive systems.

Q.5. What are the barriers in adopting energy efficiency in residential buildings? Explain.

Q.6. Explain energy conservation measures in energy efficient buildings.

# 10 Economic Analysis and Project Planning Techniques

## 10.1 LIFE CYCLE COST (LCC) AND LIFE CYCLE ASSESSMENT (LCA) METHOD

The LCC method adds for each of the alternatives of investment, the costs of acquisition, maintenance, repair, replacement, energy, and any other monetary costs, which affect the deciding factor for an investment. The value of money over time must be taken into consideration for all values, and the values are to be taken for a particular period. The measurement of all the values must be taken for a current value or net value in dollars. It is explained later on in "Discounting" and "Discount rate." For the minimum value of investment, there should be a primary reference case. Various other possibilities can be taken and compared. The one with the lowest LCC should be the investors choice of investment.

The formula shows the finding of LCCs for each alternative:

$$\text{LCC}_{A1} = I_{A1} + E_{A1} + CM_{A1} + R_{A1} - S_{A1},$$

Here, the $\text{LCC}_{A1}$ is known as the alternative life cycle cost A1, $I_{A1}$ can be said as the current value of the cost related to the investment for the alternative A1, and the current energy value is $E_{A1}$ for A1, $CM_{A1}$ is said to be the current value of the cost of operation and maintenance of the non-fuel parts of A1, the cost related to current repairing and parts substitution is represented by $R_{A1}$ for A1, and the current value related to reselling and the cost of disposal at no value for A1 is given by $S_{A1}$. The LCC method has its usage in the process of decision making. It is associated with the effectiveness of the cost. For example, in a given investment in the area of efficient energy supply renewable energy supply reduces the net cost. This can be used in the process of comparing the cost associated with the size and design of the system for a constant amount of service. If this process is correctly used, then the net cost can be reduced for the combination of investment related to efficient energy and supply of energy for a particular given facility. Unfortunately, this process cannot be used in finding which investment is best, since different investments do not give the same type of services.

### 10.1.1 LIFE CYCLE ASSESSMENT (LCA) METHOD

Life cycle assessment is the complete analysis of a product/service for the entire life cycle from an environmental impact perspective.

The environmental impact can be evaluated with the help of LCA for different products from beginning to end. There are four methodologies for LCA.

The LCA is considered a standard procedure or methodology as it provides transparency and reliability. The LCA standard is given by ISO (ISO 14044 and ISO 14040). The LCA consists of four main stages and they are as follows:

Step 1: Scope and goal
Step 2: Analysis of inventory
Step 3: Various impact assessment
Step 4: Explanation

1. **Scope and goal**: This step ensures that LCA is performing in a proper manner. An LCA models the service products or life cycle of the system. The main important parameter is to realize that the reality of the model gets distorted in some manner as the model is simplified in a complex reality. The challenges faced by LCA practitioners are to create a model in such a manner that the reality would not be affected. Hence, it is necessary to do the proper analysis and observation before performing LCA.
2. **Analysis of inventory**: In this steps, all the different environmental impacts, such as input, output, and wastages would be taken care of to get the complete analysis.
3. **Various impact assessments**: This step consists of the analysis, which makes better business plans and decisions. It is necessary to study the environmental impacts both in a favorable and adverse manner, and its effects on the organization. Also, it is necessary to give the audience a proper understanding of various impacts.
4. **Explanation**: In this phase, a detailed analysis would be done with respect to ISO standards. Also, the results are shared for proper upgradation and improvement.

## 10.2 LEVELIZED COST OF ENERGY

Levelized cost of energy (LCOE) is all the cost related to alternative investment process and the value of money over time is being taken into account for the given period of analysis. Although the process is mainly used during the comparison of supplies of alternative technologies or systems. This is different from the LCC because there is involvement of tax. However, unlike the LCC, it does not take into account the cost related to finance.

The LCOE is the value of the production of a single unit of energy which involves all cost is being returned and there is some profit associated. In equation form, this is represented as:

$$\sum\nolimits_{t-1}^{t-N} LOCE \times Q_t / \left(1+d^1\right)^t = \sum\nolimits_{t-0}^{t-N} C_t / \left(1+d\right)^t \qquad (10.1)$$

If the $d^l$ denotes the initial rate of discount, then LOCE will be given in terms of the dollar. The rate of discount (d) has its usage in bringing back the cost of the future to its present value.

## 10.3   SIMPLE PAYBACK PERIOD

Simple payback period (SPP) is the required number of years for generating the primary investment (First Cost), taking account of the total yearly saving. Its formula is displayed below:

$$\text{Simple payback period} = \frac{\text{First cost}}{\text{Yearly benefits} - \text{Yearly cost}} \qquad (10.2)$$

The SPP for a continuous deodorizer costing Rs. 60 lakhs for purchase on installment, and average of Rs. 1.5 lakhs/year for operation and maintenance will have an estimated savings of around Rs. 20 lakhs by decreasing the consumption of steam. It can be found out as follows:

Based on the payback criteria, a payback period of short duration is more required for a project:

$$\text{SPP} = \frac{60}{20 - 1.5}$$

$$\text{SPP} = 3 \text{ years } 3 \text{ months}$$

**Advantages**
This investment process has the following advantages:
- It has an easy concept and application. The less simple payback time is better for the investment.
- The projects where there is cash inflow, in the beginning, are much favored than the one where cash inflows later.

**Limitations**
- It cannot consider the value of money over time. The inflows of cash are added without taking account of the discounts during the calculation of the payback. It is a violation of the basics of financial analysis, which says that different times of cash flow can be added or subtracted after some discounts or compound.
- It does not take into account that the flow of cash after the payback period which leads to discrimination of projects that have cash inflow in the later years.

**TABLE 10.1**
**Cash Flow**

| Investment Saving in Year | Rs. 100,000 Flow of Cash A | Rs. 100,000 Flow of Cash B |
|---|---|---|
| 1 | 50,000 | 20,000 |
| 2 | 30,000 | 20,000 |
| 3 | 20,000 | 20,000 |
| 4 | 10,000 | 40,000 |
| 5 | 10,000 | 50,000 |
| 6 | — | 60,000 |

For understanding consider A and B cash flow:

The criterion of payback is much preferred for A, having the payback time of 3 years compared to B (Table 10.1),

Which is having 4 years of time for payback, although the cash inflow in B is substantial for 5 and 6 years.

- It measures that the amount of capital is recovered and not the profit.
- Even having limitations, the SPP is a more advantageous investment evaluation.

## 10.4   TIME VALUE OF MONEY

A project is associated with an investment at the initial commonly known as capital cost, and series of costs such as cost on a yearly basis and the cost related to savings for the entire lifetime of the project. In order to understand how the project is feasible, the cash flow for the present and the future must be taken into account and put on a general equation. The problem with such types of cash flow equations is the money value which undergoes changes with time. The methods related to these cash flows are called discounting, or the concept of the present value.

For instance, if money in the bank is deposited such that the rate of interest of the bank is 10% and the money deposited is Rs. 100. After a year, the money obtained will be Rs. 110 compared with the present value of the cash, which was Rs. 100.

In the same way, there is a possibility that the future value is obtained to be Rs. 90.01, for the present value of Rs. 100. If the rate of interest change, then the future value would also change. The relationship is given below:

$$\text{Value in future}\,(FV) = NPV\,(1+i)^n \text{ or } NPV = FV/(1+i)^n \qquad (10.3)$$

where:

$FV$ = Value of the future value of cash flow,
$NPV$ = Total current value of cash flow,
$i$ = Rate of interest or discount, and
$n$ = Total years in the future.

## 10.5 RETURN ON INVESTMENT (ROI)

ROI provides the "annual return" from a particular project as a percentage of the cost related to the capital. The total return includes the flow of cash through the duration of the working of the project and the rate of discount by undergoing the conversion of the net current value of the current cash flow to an amount which is equivalent to the net amount of the entire duration of the project. This can be compared to the cost of capital. The ROI does not need similar project duration and the cost associated with the capital for any comparison.

It is a significant indication of the net return which can be obtained from the initial investment of the capital, expressing in term of percent:

$$\text{ROI} = \frac{\text{Annual net cash flow first cost}}{\text{Capital cost}} \times 100 \qquad (10.4)$$

The ROI should always be more than the cost associated with the money; the more investment returns, better is the investment.

### Limitations
- No account has been taken of the value of money over time.
- There is no account of the changing nature of the annual cash inflows in this process.

## 10.6 NET PRESENT VALUE (NPV)

The net present value (NPV) for a project is the addition of all the current values for the cash flows included in it. Numerically,

$$\text{NPV} = -\frac{\text{CF}_0}{(1+k)^0} + \frac{\text{CF}_1}{(1+k)^1} + \cdots + \frac{\text{CF}_n}{(1+k)^n} = \sum_{t=0}^{n} \frac{\text{CF}_t}{(1+k)^t} \qquad (10.5)$$

where:
NPV = Total current value,
$\text{CF}_t$ = Flow of cash for the entire year "t" (t = 0,1,...n),
n = Project life, and
k = Rate of discount.

The rate of discount (k) applied for the calculation of the current value of the estimated cash flows, denoting the risk associated with the project.

$$\text{NPV}_{A1A2} \sum_{t-0}^{N} \frac{B_t - C_t}{(1+d)^t} \qquad (10.6)$$

where $\text{NPV}_{A1A2}$ is NB, i.e., current value benefits which is the total cost of the present value for the two alternatives A1 and A2 comparing both of them. Here, $B_t$ is said to be the benefits for the t year which contains the saving related to the energy and $C_t$ is the cost associated with the year t for A1 with respect to A2 and the rate of discount is d.

This method is mainly used for the long run process for finding out the profitability. The method of NB is also used for making decisions related to the investment process, and for makings designs and sizing, although this process is not beneficial for comparison of investment related to different services.

## Example

To show the NPV calculation, look at the project given below (Table 10.2):
The capital cost, K is 10% for the firm. The NPV value would be:

$$NPV = -\frac{1,000,000}{(1.10)^0} + \frac{200,000}{(1.10)^1} + \frac{200,000}{(1.10)^2} + \frac{300,000}{(1.10)^3}$$

$$+ \frac{300,000}{(1.10)^4} + \frac{350,000}{(1.10)^5} = (5273)$$

The NPV gives the overall benefit for the compensation for time and risk. Hence, the rule regarding the NPV criteria is: "Accept the project if the net present value is positive and reject the project if the net present value is negative."

### ADVANTAGES

The NPV has considerable advantages and they are as follows:
- The value of money over time is taken into account.
- The consideration of the cash flows during the project lifetime is taken into account.

**TABLE 10.2**
**Flow of Cash**

| Investment Savings (Year) | Rs. 100,000 Cash Flow |
|---|---|
| 1 | 200,000 |
| 2 | 200,000 |
| 3 | 300,000 |
| 4 | 300,000 |
| 5 | 350,000 |

## 10.7   INTERNAL RATE OF RETURN

Internal rate of return (IRR) is used for the evaluation of the amount of return which can be obtained for the investment. The IRR process displays the other possible options available in terms of return rate. The return rate which is the rate of interest for the net discounted advantages and it is equal to the net discounted costs.

Its calculation procedure is done by the method of trial and error, and the calculation of the total flow of cash is obtained for various rates of discount until the final value obtained is zero. The IRR for a project is the rate of discount, and hence the NPV becomes equal to zero. Hence, the equations become:

$$0 = \frac{CF_0}{(1+k)^0} + \frac{CF_1}{(1+k)^1} + \cdots + \frac{CF_n}{(1+k)^n} = \sum_{i=0}^{n} \frac{CF_t}{(1+k)^t}$$

where:
$CF_t$ = Flow of cash during the year ending "t,"
$k$ = Rate of discount,
$n$ = Project life.
$CF_t$ is negative for expenditure and positive for savings.

In the net present value evaluation, an assumption was made that the rate of discount is available. For the evaluation of the IRR, NPV was set to zero for finding out the rate of discount, this condition is satisfied.

For example, the calculation of IRR, take this project related to the cash flow (Table 10.3):
The IRR value is "k" and the equation satisfied is:

$$100,000 = \frac{30,000}{(1+k)^1} + \frac{30,000}{(1+k)^2} + \frac{40,000}{(1+k)^3} + \frac{30,000}{(1+k)^4}$$

**TABLE 10.3**
**Cash Flow**

| Year | 0 | 1 | 2 | 3 | 4 |
|---|---|---|---|---|---|
| Cash flow | 100,000 | 30,000 | 30,000 | 40,000 | 45,000 |

The evaluation of "k" is done by trial and error method. Many values of "k" are determined until the right side of the equation for the above case is equal to 100,000. Starting from k = 15%. The RHS becomes:

$$10,802 = \frac{30,000}{(1.15)^1} + \frac{30,000}{(1.15)^2} + \frac{40,000}{(1.15)^3} + \frac{30,000}{(1.15)^4}$$

The value obtained is more than 100,000. Hence, k value is incremented from 15% to 16%. The RHS becomes:

$$98,641 = \frac{30,000}{(1.16)^1} + \frac{30,000}{(1.16)^2} + \frac{40,000}{(1.16)^3} + \frac{30,000}{(1.16)^4}$$

Now the value of NPV is less than 100,000; hence, it can be said that of k lies between 15% and 16%. For most of the cases, it is applicable.

### Advantages
The IRR criteria has a few advantages:
- Consideration of the value of money over time.
- Consideration of the cash flow stream throughout.
- It is sensible for businessmen to think in terms of return rate and find a significant value, like NPV, which is complicated in working.

### Limitations
- The value of the IRR cannot differentiate the process of lending and borrowing, and so more IRR is not necessarily of more use.

### Example

Find the value of a IRR for economizer costing Rs. 500,000, lasting 10 years, resulting in savings of fuel of Rs. 150,000 yearly. Find i equating the following:

$$\text{Rs.}500,000 = 150,000 \times PV(A = 10 \text{ years}, i = ?)$$

To solve this find (NPV) for different values of i, selected by manual assumption;

$$\text{NPV } 25\% = \text{Rs.}150,000 \times 3.571 - \text{Rs.}500,000$$
$$= \text{Rs.}35,650$$

$$\text{NPV } 30\% = \text{Rs.}150,000 \times 3.092 - \text{Rs.}500,000$$
$$= -\text{Rs.}36,200$$

For i = 25%, NPV is positive; i = 30%, NPV is negative. Thus, for some discount rates between 25% and 30%, PV benefits = cost of PV. To obtain the rate more accurately, one will be able to find interpolation among the two rates as follows:

$$i = 0.25 + (0.30 - 0.25) \times 35,650/(35,650 + 36,200)$$
$$= 0.275, \text{ or } 27.5\%$$

## 10.8 CASH FLOWS

The cash flow is generally of two types; the investment is initially having one or more than one installment and an investment savings. This concept is straightforward, based on the management of energy investment reality. For a project, other forms of cash flows are also present, which are:

- Cost related to capital such as design, planning, installation and project commission, is a one-time cost, and hence, inflation or discount rate factors does not affect it.
- Net flow of cash such as savings obtained yearly, obtained every year throughout the period of project work which includes types of equipment, insurance, equipment leases, cost of energy, etc., if there is any increment related to any such costs your cash flow will be of negative decrement and increment of cost is a positive flow of cash. Factors associated with the calculation of net flow cash are shown in Figure 10.1.
- Taxes, using the tax rate of marginal value are applicable for the positive or negative cash flows.
- Asset depreciation, the depreciation of plant assets over their life; depreciation is a "paper expense allocation" rather than cash flow of reality, and

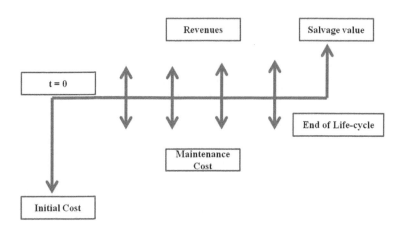

**FIGURE 10.1**   Cash flow diagram.

hence, is directly present in the life cycle cost. But depreciation is a "real expense" according to evaluation of tax, and hence has an effect on the tax evaluation given above.

- Intermediate flow of cash occurs under sudden occasions rather than throughout the year during the project life.

## 10.9   SENSITIVITY ANALYSIS

The factor of uncertainty is the integral part of the many cash flows. However, in some cases, the exact estimation of cash flow is possible like capital cost, energy cost savings, and maintenance costs. The factor of inflation plays an important role to create uncertainty in the future cash flow. Sensitivity analysis is an analysis of the danger or risk in a project. Due to uncertainty, sensitivity analysis (SA) became an important analysis in a project. Now SA becomes very easy because of computer spreadsheets which are splendidly available in free of cost basis. This can be easily customized based on project requirements. The whole objective of this analysis is to provide an enhanced project design with clear action against probable future problems.

There are few major and minor factors to be considered very carefully to do this analysis in better way. These are the major factors namely operational cost, capital cost, debt from bank/other sources, shifting the mode of operation of finance like leasing and changing the project period. The minor factors are to reduce/increase in interest/tax rate, changes in the auditing standards, reduce/increase in depreciation rates, extension of government funded project, change in service rule, energy cost, and change in technology.

## 10.10   PROJECT PLANNING TECHNIQUES; CPM AND PERT

One of the most challenging jobs as a manager is to handle a project of considerable size in proper coordination of the activities in the entire organization. A vast area of detailing is to be considered for the planning and proper coordination of the activities. For the development of a schedule that is realistic in nature and proper look into the progress related to the project.

Although two techniques of operation research (OR) are PERT (program evaluation and review technique) and CPM (critical path method). These are generally used by the project manager for proper handling of the responsibilities. These two methods have high usage of networks for planning, helping and displaying the activities based on the coordination. It uses software for handling such data, developing the information and the monitoring of the project progress.

Software for the management of a project, for example, MS project for OR is now available for such purpose. The PERT and CPM have been used for projects of large varieties.

An example will illustrate the PERT/CPM technique.

## Example

The optimistic, most probable, and pessimistic times (in days) for completion of activities for a specific project are as follows (Table 10.4):
1. Find the critical path.
2. Find the probability that all critical activities will be completed in 35 days or less.

SOLUTION
1. The expected time is obtained by using the formula $(O + 4M + P)/6$ and the variances are obtained by using the formula $((P - O)/6)^2$ and are summarized in the following Table 10.5:

**TABLE 10.4**
**The Optimistic, Most Probable, and Pessimistic Times (in Days)**

| Activity | Immediate Predecessor | Optimistic Time (O) | Most Probable Time (M) | Pessimistic Time (P) |
|----------|----------------------|--------------------|------------------------|----------------------|
| A | — | 4 | 5 | 6 |
| B | — | 6 | 8 | 10 |
| C | A | 6 | 6 | 6 |
| D | B | 3 | 4 | 5 |
| E | B | 2 | 3 | 4 |
| F | C, D | 8 | 10 | 12 |
| G | E | 6 | 7 | 8 |
| H | C, D | 12 | 13 | 20 |
| I | F, G | 10 | 12 | 14 |

**TABLE 10.5**
**Expected Times and Variances**

| Activity | Expected Time | Variance |
|----------|---------------|----------|
| A | 5 | 0.11 |
| B | 8 | 0.44 |
| C | 6 | 0 |
| D | 4 | 0.11 |
| E | 3 | 0.11 |
| F | 10 | 0.44 |
| G | 7 | 0.11 |
| H | 14 | 1.78 |
| I | 12 | 0.44 |

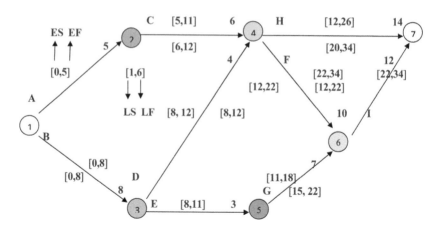

**FIGURE 10.2**   The numbers in brackets above the arrows show the earliest start (ES) and earliest finish (EF) for each activity, respectively, whereas the numbers in brackets below the arrows show the last start (LS) and last finish (LF) times, respectively.

The project network diagram based on the expected values in Table 10.5, is depicted below (Figure 10.2):

$$EF = ES + Activity\ duration$$

and

$$LF = LS + Activity\ duration$$

$$Activity\ slack = LS - ES = LF - EF$$

Activity slack shows how much an activity can be delayed without affecting the project completion time.

For example, in our problem slack for activity E = 12−8 = 4; hence, we can delay activity E for 4 days and still finish the project in the expected completion time of 34 days.

On the other hand, when the activity slack is zero, that activity can be delayed zero-days, meaning that it cannot be delayed; hence, an activity is critical whenever its slack is zero. Such an activity is called critical because any delay in the completion of that activity would delay the whole project. In our example, activities B, D, F, and I there is zero slack; hence, they are critical activities. All other activities, in our example, are non-critical. Knowing which activities are critical and which are non-critical can be quite useful; whenever there is an unexpected delay in critical activities we can shift resources like capital, workforce, etc., from non-critical activities that can be delayed to the critical activities which cannot be delayed.

In our problem, the critical path can be depicted as follows:

$$Critical\ path : B \rightarrow D \rightarrow F \rightarrow I\ or\ 1 \rightarrow 3 \rightarrow 4 \rightarrow 6 \rightarrow 7$$

Expected completion time, $\mu$ = 34 days
   Variance along the critical path:

$$\sigma^2 = \sigma^2_B + \sigma^2_D + \sigma^2_F + \sigma^2_I = 0.44 + 0.11 + 0.44 + 0.44 = 1.43$$

$$\sigma = \sqrt{1.43} = 1.2$$

2. $P(x \le 35) = P\left(z \le (35-34)/1.2\right) = P(z \le 0.83) = 0.7967 \approx 0.8 = 80\%$
   NOTE: $z = (x-\mu)/\sigma$ for standard normal distribution. In our problem, $x = 35$, $\mu = 34$, and $\sigma = 1.2$ and that is how we get P(z $\le$ (35−34)/1.2) above.

   The PERT/CPM also gives managers a decent idea about the probability of project completion in order to create further planning or make a few of the processes faster if needed. Also, these probability computations help managers to select and accept those projects that have higher chances of completion within a specified time frame.

## QUESTIONS

Q.1.  What do you mean by the payback period?

Q.2.  A cogeneration plant installation was to reduce the net yearly energy bill by Rs. 24 lakhs. Now the capital cost for the new cogeneration installation is Rs. 90 lakhs, and the net cost of maintaining and operating are Rs. 6 lakhs. Find the expected payback duration for the project.

Q.3.  A heat exchanger cost is Rs. 1 lakh. Find the simple payback duration taking into account the possibility of annual savings and operating of Rs. 60,000/- and Rs. 15,000/-, respectively.

Q.4.  Find the simple payback time for a boiler of Rs. 75 lakhs during purchase and an average of Rs. 5 lakhs yearly for operation and maintenance and is expected to save Rs. 30 lakhs annually.

Q.5.  Explain the main drawback of the simple payback method.

Q.6.  Illustrate the advantages of different methods of simple payback.

Q.7.  Explain the term "present value of money."

Q.8.  Explain the term "discounting."

Q.9.  Explain ROI.

Q.10.  The energy proposal investment is Rs. 10 lakhs. Yearly savings for the initial three years is 150,000, 200,000, and 300,000. Taking capital cost as 10%, find the total present value of the investment.

Q.11.  What are the advantages of the net present value?

Q.12.  What are the advantages and limitations of the method of discounted cash flow?

Q.13.  Explain the idea behind carrying out sensitivity analysis?

# References

Bai, Yuxing, et al. "Effect of blade wrap angle in hydraulic turbine with forward-curved blades." *International Journal of Hydrogen Energy* 42(29) (2017): 18709–18717.

Birol, F. *Key World Energy Statistics*. International Energy Agency, 2017.

Brunet, Robert, Gonzalo Guillén-Gosálbez, and Laureano Jiménez. "Combined simulation–optimization methodology to reduce the environmental impact of pharmaceutical processes: Application to the production of Penicillin V." *Journal of Cleaner Production* 76 (2014): 55–63.

Chevula, Sreenadh, Ángel Sanz-Andres, and Sebastián Franchini. "Estimation of the correction term of pitot tube measurements in unsteady (gusty) flows." *Flow Measurement and Instrumentation* 46 (2015): 179–188.

Glowacz, Adam. "Acoustic based fault diagnosis of three-phase induction motor." *Applied Acoustics* 137 (2018): 82–89.

Haque, M. E., M. S. Islam, M. R. Islam, H. Haniu, and M. S. Akhter. "Energy efficiency improvement of submersible pumps using in barind area of Bangladesh." *Energy Procedia* 160 (2019): 123–130.

Hossain, Altab, et al. "An intelligent flow control system of coolant for a water reactor based cooling tower." *Energy Procedia* 160 (2019): 566–573.

Jung, Jae Hyuk, and Won-Gu Joo. "The effect of the entrance hub geometry on the efficiency in an axial flow fan." *International Journal of Refrigeration* 101 (2019): 90–97.

Li, Qing, et al. "Dynamic simulation and experimental validation of an open air receiver and a thermal energy storage system for solar thermal power plant." *Applied Energy* 178 (2016): 281–293.

Prithvi Raj, P., P. Suresh Kumar Gupta, P. Praveen Kumar, G. Paramesh, and K. Rohit. "Model and analysis of rotor and impeller of eight stage centrifugal pump." *Materials Today: Proceedings* 21 (2020): 175–183. https://doi.org/10.1016/j.matpr.2019.04.215

Saidur, Rahman, Jamal Uddin Ahamed, and Haji Hassan Masjuki. "Energy, exergy and economic analysis of industrial boilers." *Energy Policy* 38(5) (2010): 2188–2197.

Xu, Chen, and Yijun Mao. "Experimental investigation of metal foam for controlling centrifugal fan noise." *Applied Acoustics* 104 (2016): 182–192.

Yu, Xiaobing, and Yansong Shen. "Model study of central coke charging on ironmaking blast furnace performance: Effects of charing pattern and nut coke." *Powder Technology* 361 (2020): 124–135. https://doi.org/10.1016/j.powtec.2019.10.012

# Index

**A**

air humidity, 144
ambient temperature, 144
ammeter, 57
analysis of data and information, 97
attractive, 15
auditor, 55
audit phase, 43
automatic blowdown control, 127
axial flow fans, 113

**B**

batteries, 22
Bhabha Atomic Research Centre (BARC), 10
block rate tariff, 16
blowers, 28
boiler, 123
    performance evaluation, 124
    types and classification, 124
broad stakeholder participation, 150
building(s), 32
    envelope, 152
    heat loads, 36

**C**

calculate CUSUM, 102
capacitor locations, 106
carbon monoxide, 58
cash flows, 167
CFL, 34
chillers, 30
clamp meter, 58
cold insulation, 36
combined heat and power (CHP), 135
commercial customers, 16
compressed air, 29
compressed-air systems, 155
compressed natural gas (CNG), 11
compressors, 29
computers, 21
continuous re-circulating bogie type furnaces, 133
continuous steel reheating furnaces, 129
conventional ballasts, 36
cooking, 22
cooling towers, 31
core losses, 119

crude oil, 8
cumulative sum control chart (CUSUM), 99
    graph, 103

**D**

damper controls, 116
day lighting, 153
detailed energy audit, 40
determine baseline relationship, 99
DG sets, 32
direct efficiency, 125
drives, 27
drying, 22

**E**

E.C.B.C. development, 149
economizer, 139
    for pre-heating feed water, 126
efficiency of fan, 115
efficiency of pump, 110
electrical utilities, 27
electricity distribution system, 27
elements of targeting & monitoring system, 97
energy conservation opportunities in fans, 115
energy distribution, 47
energy-efficient motors, 117
energy manager, 84
energy performance model, 98
energy-saving, 42
enthalpy, 71
established economy, 18
evaporation ratio, 125
excess control of air, 126

**F**

fairness, 15
fans, 28
    systems, 111
filament lamps, 34
fire-tube boilers, 124
fixing, bubbling & fast fluidized beds, 134
flat rate tariff, 16
fluidized bed combustion (FBC) boilers, 124, 134
friction and windage losses, 119
furnaces, 26, 128
    heat transfer, 130

## G

gas turbine cogeneration systems, 136

## H

heating and cooling systems, 20
heat loss due to radiation and convection, 127
high-frequency electronic ballasts, 36
high-pressure sodium vapor (HPSV) lamps, 34
H.V.A.C. A.C. systems, 154
hydrogen, 58

## I

identify metrics, 87
improvement of power factor, 106
incandescent lamps, 34
incomplete combustion, 126
industrial heating system, 128
inlet guide vanes, 116
insulation, 26
internal rate of return (IRR), 165
International Energy Agency (IEA), 3

## K

kVA maximum demand tariff, 17
kW and kVAr tariff, 18

## L

large purchasing decisions, 22
latent heat, 71
LED panel indicators, 34
levelized cost of energy (LCOE), 160
life cycle assessment (LCA), 159
life cycle cost (LCC), 159
lighting, 153
    system, 19
load curve, 105
load management strategies, 105

## M

major ECOs, 19
maximum demand basics, 105
maximum demand tariff, 17
medium ECOs, 19
metal halide lamps, 34
microwave ovens and electric kettles, 21
minor ECOs, 19
motors, 27, 154
    speed, 117

## N

National Disaster Management Authority
    (NDMA), 10
National Disaster Response Force (NDRF), 10
net present value (NPV), 163
net pressure, 113
New Policies Scenario (NPS), 3
normalize data, 86

## O

oil fired furnace, 128

## P

packaged boilers, 124
Pitot tube, 113
Planning Commission of India, 7
plant energy performance (PEP), 51
plant load factor (PLF), 5
plate heat exchanger, 140
positive displacement, 109
post-audit phase, 44
power factor tariff, 17
pre-audit phase, 43
precipitation, 145
preheating of combustion air, 126
process heat loads, 36
products storage in material processing utilities,
    105
project planning techniques, CPM and PERT, 168
propeller, 113
proper return, 15
publish results, 87
pulley change, 116
pump, 28, 107

## Q

qualitative appraisals, 88
quantitative appraisals, 88

## R

reasonable profit, 15
recuperators, 139
reduction of boiler steam pressure, 127
refrigeration load, 75
regenerator, 139
regression and CUSUM, 103
rerolling mill furnace, 128
rescheduling and staggering loads, 105
return on investment (ROI), 163

room air conditioners, 20
rotary hearth furnace, 132
rotodynamic pumps, 107–108

**S**

shell and tube heat exchanger, 140
simple payback period (SPP), 161
simple tariff, 15
simplicity, 15
sky condition, 147
sliding scale tariff, 17
social configurations, 18
sodium lamps, 34
solar radiation, 144
stack temperature, 126
standard energy audit, 40
static pressure, 113
stator and rotor losses, 118
steam systems, 25
steam turbine cogeneration systems, 135
stray load-losses, 119
summer opportunities, 23
synchronous speed, 117

**T**

thermal efficiency, 125
thermal insulators, 140
thermal utilities, 24
thermocompressor, 140
three-part tariff, 18
time value of money, 162

Toray Textiles, 96
tracking of data, 86
transmission and distribution (T&D), 5
tube-axial, 113
two-part tariff, 16

**U**

utility cost analysis audit, 40
utility data analysis, 40

**V**

vane-axial, 113
variable speed drives, 116
velocity pressure, 113
    velocity calculation, 114
ventilation and indoor air quality, 152
voltmeter, 57

**W**

walking beam furnace, 133
walking hearth furnaces, 131
walk-through audit, 40
walk-through survey, 40
waste heat recovery, 26, 138
waste water, 33
water heaters, 21
water-tube boiler, 124
wind, 147
winter opportunities, 23